PRINCIPLES OF ANIMAL TAXONOMY

NUMBER XX OF THE

Columbia Biological Series

EDITED AT COLUMBIA UNIVERSITY

GEORGE GAYLORD SIMPSON

The Museum of Comparative Zoology

Harvard University

Principles of Animal Taxonomy

New York and London

Columbia University Press

SBN 231–02427–4
Library of Congress Catalog Card Number: 60-13939
Copyright © 1961 Columbia University Press
First printing 1961
Fourth printing 1969
Printed in the United States of America

Columbia Biological Series

EDITED AT COLUMBIA UNIVERSITY

Preface

> Is it not extraordinary that young taxonomists are trained like performing monkeys, almost wholly by imitation, and that in only the rarest cases are they given any instruction in taxonomic theory?—A. J. CAIN.

There are authorities who maintain that one should not think much about taxonomy but just do it. I have been doing it for many years, but like the would-be philosopher who was thwarted by having cheerfulness break in, I have found thought breaking in. In 1941–42 (published in 1945) I put down some of those thoughts in connection with my broadest taxonomic effort, a classification of the Mammalia. That essay was intended to explain and justify the bases of that particular classification, but it also seemed to have some wider interest for taxonomists. Ever since then the divorcing of the essay from the classification of mammals and its separate issue in revised form have been one of those projects to be performed "when I have time."

One of the advantages of lecture series—their greatest advantage, I believe—is that they may force the lecturer to take the time to organize his thoughts and to write them down. The flattering invitation to join the roster of Jesup Lecturers at Columbia University was accepted, somewhat rashly, in large part because it was a means of forcing myself to take the time for the long-deferred project on the principles of taxonomy. Of course this turned out to involve much more than just dusting off the old essay, revising and expanding it. Indeed the present book, based on the Jesup Lectures for 1960, has little to do with the earlier essay aside from the facts that both are on the same subject and that I still agree with much, by no means with all, that I thought fifteen to twenty years ago.

Here I have tried something much more ambitious than in the earlier essay. As far as is permitted by ability and scope, I have here tried to examine the deepest foundations of taxonomy and to build up from those foundations the structures of zoological classification. Further characterization of the subject matter is made at the beginning of Chapter 1, where it is more likely to be read, and the book before you demonstrates for itself what has been made of the subject. I might here add that, although the treatment is not intended to be elementary, I have inserted much that professional taxonomists already know and may, for their purposes, find superfluous. (Various of those passages were omitted from the lectures.) That material is included with the hope that students may here acquire some of the rudiments while also, and perhaps with greater difficulty, learning to think about taxonomy and not just to do it.

Dr. Anne Roe has read the entire manuscript and has helped me to limit, if not altogether to eliminate, failures of communication. She deserves more than the routine acknowledgment often given to wives. Miss Holly Osler has gone beyond the strict call of duty in converting my handwriting into legible manuscript and in assisting with details of bibliography and index. At Columbia University Press, Raymond J. Dixon has edited the manuscript and Miss Nancy Dixon has designed the book. Mrs. Nancy Gahan has converted my roughs into finished illustrations.

I am grateful, perhaps most of all, to the late Alexander Agassiz and to Harvard University. By a conjunction foreseen by neither one, they have given me the freedom to pursue these thoughts to this point.

GEORGE GAYLORD SIMPSON

Cambridge
April, 1960

Contents

Figures

PRINCIPLES OF ANIMAL TAXONOMY

1

Systematics, Taxonomy, Classification, Nomenclature

Any discussion should start with a clear understanding as to what is to be discussed. A main purpose of this chapter, therefore, is to establish precisely what is meant by taxonomy in this book. That also involves consideration of systematics, which is broader than taxonomy and includes it, and of classification and nomenclature, which are narrower than taxonomy and are (in a somewhat different way) included in it. A second prerequisite of useful discussion is the designation or, if need be, the establishment of a vocabulary. This chapter therefore includes some definitions and explanations of a general terminology that will be used throughout the later chapters. It is further useful at this point to mention certain general principles that underlie the whole subject, some of which will be exemplified more concretely and developed in further detail in later stages of the enquiry. Such preliminaries also involve brief mention of certain problems, tentative conclusions, questions of procedure, and other debatable points that, again, are to be explored more fully and systematically as the discussion proceeds. Finally some consideration of the formal framework of classification and the conventions of nomenclature is required background for all that follows.

One point is to be made before even beginning the discussion: the title of this book is to be taken literally. The subject is principles, and particular groups of organisms or classifications thereof are involved only as examples and as the basis for inductive derivation of prin-

ciples. The principles are those of taxonomy, as strictly defined here-
inafter. Broader principles of systematics in general are brought in
only as far as necessary to provide a background for or to elucidate the
origin of specifically taxonomic principles. The kind of taxonomy involved is explicitly that applicable to animals, by which I mean metazoans. Most of the same principles also apply to plants and many of
them, with less generality, to protists. Although such applications may
be mentioned in passing, they are largely incidental. Examples are
drawn principally from the mammals, for personal and practical reasons. Principles derived from and applicable to mammals are sufficiently general for all Metazoa that the word "animal" in the title
should not constitute false advertising, but a specialist in, say, coelen-
terates may quite properly reach some other conclusions in the light
of his different experience. Important differences of opinion are discussed, but I have not tried to abstract the enormous literature of the
subject and still less to produce a neutral expression of a consensus.

ON ORDER IN NATURE AND ORDERING IN SCIENCE

"The world is so full of a number of things"—that it can be extremely, even hopelessly, confusing. If each of the many things in the
world were taken as distinct, unique, a thing in itself unrelated to any
other thing, perception of the world would disintegrate into complete
meaninglessness. That would be true if each thing, say for instance
each tree, were considered as a wholly separate individual; of course
it would not then *be* a tree, for "tree" is a collective concept not applicable to a single object considered without relationship to any other.
It could indeed be argued (and it has been argued at length) that
even the individual tree could not be perceived as a thing in itself, for
each of the sensory perceptions is meaningful only as a collective
generalization, and without some form of ordering and abstraction the
tree would disintegrate into a formless mosaic of "green" (distinct from
the sensation derived from any other green tree), "rough feeling,"
"branching," and so on.

There could be no intelligible language if each thing (or each
perception) were designated by a separate word, and no rational
thought if symbols did not generalize characteristics and relationships

shared by innumerable different objects. The necessity for aggregating things (or what is operationally equivalent, the sensations received from them) into classes is a completely general characteristic of living things. One should hesitate, although some do not, to apply such words as "consciousness" or "perception" to an amoeba, for instance, but it is perfectly obvious from the reactions of an amoeba that something in its organization performs acts of generalization. It does not react to each bit of food, say, as a unique object, but in some way, in some sense of the word, it *classifies* innumerable different objects all within the class of foodstuffs. Such generalization, such classification in that sense, is an absolute, minimal requirement of adaptation, which in turn is an absolute and minimal requirement of being or staying alive.

The relationships among things that must in some way be taken into account are of many different kinds and are themselves intricately interrelated. Among these are two that can usually be distinguished and that seem to be particularly fundamental. They nearly correspond with what the psychologists used to call "association by contiguity" and "association by similarity," and we may here borrow those terms and bend them somewhat to our different purposes.[1] Association by contiguity (for our purposes) is a structural and functional relationship among things that, in a different psychological terminology, enter into a single *Gestalt*. The things involved may be quite dissimilar, or in any event their similarity is irrelevant. Such, for instance, is the relationship between a plant and the soil in which it grows, between a rabbit and the fox that pursues it, between the separate organs that compose an organism, among all the trees of a forest, or among all the descendants of a given population. Things in this relationship to each other belong both structurally and functionally to what may be defined in a broad but technical sense as a single system.

Association by similarity is a more nearly self-defining concept, if only because this is itself the basis of the linguistic or other symbolic systems that we use in framing definitions. It means, of course, simply the classing together of various different things because all possess some one or more common characteristics. The objects in such groups

---

[1] The terms were originally used in now outmoded forms of learning theory. They are here applied directly to perception and implicitly to relationships that are assumed (or postulated) to occur objectively in the environment.

are by definition similar in some degree, but any structural or functional relationships among them are irrelevant and they do not ordinarily form a system in the sense usually adopted in modern system theory. For instance, they do not interact, or it is no part of *this* relationship if they do. They may be all yellow, or all smooth, or all with wings, or all ten feet high. They may, for that matter, all be seats with four legs and a back, in which case we call them "chairs"—an example of the fact that the words of a language (with the trivial exception of proper nouns) all embody associations by similarity.

Evidently in some instances the same relationship may be considered alternatively or concomitantly as contiguous or similar. For example, the class of all descendants of a given population, already given as an instance of association by contiguity, may also be considered as associated by similarity of ancestry. That is stretching a point, because the similarity is not *within* the objects so classed. In practice, as we all know, there will in this situation also be similarities that *are* within them, and hence an association by similarity that is concomitant with or produced by but not identical with the association by contiguity. The example is obviously relevant to taxonomy and indicates that these basic and abstract considerations are fundamental to our subject.

As is true of many characteristics of living things, perceptions of associations by contiguity and by similarity have reached unparalleled heights of intensity and diversity in the human species. We certainly order our perceptions of the external world more fully, more consistently, and more consciously than do any other organisms, and we usually order them in one or indeed in both of these two ways. Such ordering is most conspicuous in the two most exclusively human and in some sense highest of all our activities: the arts and the sciences. It is this intensity and completeness of ordering, and of ordering in the same two fundamental ways, that the sciences and the arts have in common. It is also, I believe, this degree and these kinds of ordering that are to be considered as truly aesthetic (in both arts and sciences). It is pertinent here to insist that taxonomy, which is ordering par excellence, has eminent aesthetic value. Unfortunately, however, our enquiry specifically into taxonomy would go astray if we pursued this fascinating point further in this place.

It has been pointed out that scientists, even more than most mortals, must tolerate uncertainty and frustration (Roe, 1946). It has become increasingly evident in our century that science is uncertain in its very nature. With exceptions mostly on a trivial and strictly observational level, its results are rarely absolute but usually establish only levels of probability or, in stricter terminology, of confidence. Scientists must also tolerate frustration because they can never tell beforehand whether their operations, which may consume years or a lifetime, will generate a desired degree of confidence. (If this could be told beforehand, the operations would be unnecessary.) Indeed one thing of which scientists can be quite certain is that they will not achieve a *complete* solution of any worth-while problem.

Scientists do tolerate uncertainty and frustration, because they must. The one thing that they do not and must not tolerate is disorder.[2] The whole aim of theoretical science is to carry to the highest possible and conscious degree the perceptual reduction of chaos that began in so lowly and (in all probability) unconscious a way with the origin of life. In specific instances it can well be questioned whether the order so achieved is an objective characteristic of the phenomena or is an artifact constructed by the scientist. That question comes up time after time in animal taxonomy, and it will come up repeatedly in various guises in the following pages. Nevertheless, the most basic postulate of science is that nature itself is orderly. In taxonomy as in other sciences the aim is that the ordering of science shall approximate or in some estimable way reflect the order of nature. All theoretical science is ordering and if, as will soon be discussed, systematics is equated with ordering, then systematics is synonymous with theoretical science. Taxonomy, in any case, is a science that is most explicitly and exclusively devoted to the ordering of complex data, and in this respect it has a special, a particularly aesthetic (as has been said), and (as might be said) almost a superscientific place among the sciences.

[2] It is not a real contradiction that the most creative scientists are frequently just those not only willing to accept the existence of disorder but also positively attracted to it. The evident reason is that the recognition of disorder is an opportunity and in fact a necessary preliminary for the creative act of ordering.

THE ROLES OF SYSTEMATICS, TAXONOMY, AND CLASSIFICATION

The idea that all science involves ordering and that systematics in that general sense is therefore coextensive with science has been stressed by Hennig (1950) in the following passage:

In order now to be able to judge correctly the position of systematics in the field of biology and the role that it is called upon to play in the solution of the basic problems of this science, one must first make clear that there is a systematics not only in biology, but that it is rather an integrating part of any science whatever. It is surprising and peculiar to see to what degree the original significance of this concept has been forgotten in biology in the course of the fundamentally inadmissible but now general limitation of the concept of systematics to a particular subdivision of the science as a whole.[3]

That conception of systematics is etymologically justifiable and quite rightly emphasizes the connection between biological systematics and the ordering of science in general. It also rightly objects to equating systematics to so narrow a scope as, for instance, classification. Nevertheless it is contrary to current usage and it is terminologically awkward, because we do not need another term for theoretical science overall and we do need a term for broad but special aspects of ordering essential in biology. Unfortunately it appears that even in biology there is as yet no strong consensus as to the definition of the term "systematics." It is universally agreed that systematics includes the formal, technical classification of organisms. At one extreme the two are equated, and systematics is defined, simply, as the science of classification. It is, however, evident that even those who give that narrow definition actually include in their discussions a great deal more than classification in any strict sense. (See, for example, Mayr, Linsley, and Usinger, 1953, and especially Mayr, 1942, where "sys-

---

[3] "Um nun die Stellung der Systematik im Rahmen der Biologie und die Rolle, die sie bei der Lösung der Grundfragen dieser Wissenschaft zu spielen berufen ist, richtig beurteilen zu können, muss man sich zunächst klarmachen, dass es 'Systematik' nicht nur in der Biologie gibt, dass sie vielmehr einen integrierenden Bestandteil jeder Wissenschaft überhaupt bildet. Es ist erstaunlich und befremdlich, zu sehen, wie sehr man in der Biologie über der im Grunde genommen unzulässigen aber heute allgemein üblichen Einschränkung des Begriffes der Systematik auf ein bestimmtes Teilgebiet der Gesamtwissenschaft die Bedeutung vergessen hat, die diesem Begriff ursprünglich zukommt."

tematics" is defined only as equivalent to taxonomy but where both terms are consistently used in a sense much broader than classification.) Indeed there is a strong tendency among biologists to give systematics a somewhat vague extension well beyond the more definite, relatively narrow bounds of classification. This tendency has been authoritatively exemplified by Blackwelder and Boyden (1952): "Systematics is the entire field dealing with the kinds of animals, their distinction, classification, and evolution." In keeping with this tendency and much modern usage, an even broader but equally clear definition may be adopted:

Systematics is the scientific study of the kinds and diversity of organisms and of any and all relationships among them.

In this definition, the word "relationships" is to be understood not in any particular, narrow sense (for instance, in the sense of phylogenetic connections), but in a fully general way, including specifically all associations of contiguity and of similarity as discussed in the last section.

Classification of organisms is an activity that belongs exclusively to systematics, but this exclusive relationship does not warrant equation of the two. In its relationship to other biological specialties, systematics is a focus of interest or a way of looking at things that may derive from or apply to almost any sort of biological study. For example, molecular biology does not necessarily or (at present) even usually focus attention on the kinds or diversity of organisms, and therefore as a special biological science it is not a part of systematics. Nevertheless, the data of molecular biology, when comparative at any level (either within or between populations), most decidedly enter into systematics—the distribution of the various hemoglobins is only one of innumerable strikingly pertinent examples. Systematics, in turn, does apply to the whole field of molecular biology and supplies one of the several ways in which the results of that subject may be explained or meaningfully ordered. There is an analogy with the subject of evolution, which (beyond, perhaps, phylogeny) has little subject matter that belongs *exclusively* to it, but which draws data from all the biological specialties and which in turn provides interpretive and explanatory principles applicable to all. As I pointed out years ago (Simpson, 1945):

[Systematics] [4] is at the same time the most elementary and the most inclusive part of zoology, most elementary because animals cannot be discussed or treated in a scientific way until some [systematization] [4] has been achieved, and most inclusive because [systematics] [4] in its various guises and branches eventually gathers together, utilizes, summarizes, and implements everything that is known about animals, whether morphological, physiological, psychological, or ecological.

It is evident that all comparative biological sciences enter into systematics, because they necessarily study relationships among diverse organisms. That is notably true of comparative anatomy, comparative physiology, and comparative psychology—all in their broadest senses and thus including, for instance, comparative cytology, comparative biochemistry, and ethology (one approach to the comparative study of behavior). The systematic aspects of those sciences involve in large part—although by no means exclusively—associations of similarity. In some other sciences the associations are mostly those of contiguity, and then the pertinence to systematics, although no less real and important, may not be so obvious. That is particularly true of biogeography and ecology.

The pertinence of biogeography to systematics is generally recognized, implicitly at least, and requires little discussion here. For example, Mayr's *Systematics and the Origin of Species* (1942) employs zoogeographical data throughout, and Darlington's *Zoogeography: the Geographical Distribution of Animals* (1957) is at least as much a work on systematics as on geography. The pertinence of ecology is less generally noted and has, indeed, been denied by some systematists. Thus Blackwelder and Boyden (1952), in the paper previously quoted, took a then exceptionally broad view of systematics but nevertheless excluded ecology, and one of the founders of "the new systematics," Thorpe (1940) maintained that:

[Synecology] has in itself had relatively little effect on the studies of systematists beyond giving them a great deal of identification work and taking much valuable time and energy which might in many cases have been more usefully employed.

Synecology is the study of multispecific communities, as opposed to autecology which is the study of the relationships between partic-

---

[4] The word in the original is "taxonomy," which I then used as I here use "systematics." As discussed below, I now prefer a different usage.

ular kinds of organisms and their environments. It would appear that in reality synecology is not only pertinent to but is a branch of systematics as here conceived, because it is the study of some sorts of relationships (various associations of contiguity) among diverse organisms. Autecology, to be sure, is less directly connected with systematics, but when studied comparatively it plays the same role as other comparative biological studies in providing data for systematics, mostly in the form of associations of similarity. It is, for example, essential in the study of convergence, which, as will later be exemplified, is one of the key problems in systematics.

Although this fact is really obvious, it should further be made explicit that genetics is particularly closely allied to systematics. The species problem, one of the major foci of attention in systematics, is fundamentally a genetic problem. Genetics is no less extensively and intimately involved in an enormous range of other problems in systematics, such as those of morphogenesis (both ontogenetic and phylogenetic), variation, isolation, population dynamics, and many more, various of which will be touched on later in this study.

The activity of systematics in studying relationships (in the broadest sense) among kinds of organisms almost necessarily requires the erection of some formal system in which the organisms in question are grouped into classes and of a vocabulary by which those classes are designated. In other words, as has already been mentioned without defining the terms, it involves classification and nomenclature, which for present purposes may be defined as follows:

Zoological classification is the ordering of animals into groups (or sets) on the basis of their relationships, that is, of associations by contiguity, similarity, or both.

Zoological nomenclature is the application of distinctive names to each of the groups recognized in any given zoological classification. Nomenclature is thus an essential adjunct or secondary outcome of classification. (In principle the "names" of this definition need not necessarily be words but could be almost any kind of symbol. Although other kinds of symbols have occasionally been proposed for special purposes, in practice they almost always are words and that usage may be assumed in the present enquiry.)

It should by now be sufficiently clear that although classification is an essential part of systematics and is the biological activity most exclusive to that field, it composes only a comparatively narrow part, far from the whole of systematics. Even with this definition, which is sharper than usual, the word "classification" remains somewhat ambiguous. For one thing, like many analogous terms (for example, adaptation), it is commonly applied both to a process, the classifying of organisms, and to a result of that process, an explicit arrangement of the names applied to the classes in which particular, specified organisms have been grouped. The distinction is real and important, but at present I see no reason for consistent use of two different terms because the meaning, as between these two, is usually evident in context. Use of one term with both meanings rarely induces logical confusion.

There is, however, another distinction that must be clearly made if confusion is to be avoided. That distinction is between the process of classification and the methods, procedures, rules, and principles in accordance with which that process is carried out. This is nearly the distinction forcefully made by Gregg (1954) between what he calls "taxonomy proper," which is essentially the process of classification, and "methodological taxonomy." Gregg explains that "the primary domain of inquiry for taxonomy proper is the organic world, whereas the domain of methodological taxonomy is taxonomy proper itself." The distinction is essential and should certainly be maintained in terminology, but these particular terms seem to me clumsy if not confusing.

"Taxonomy proper" is simply one of the time-hallowed meanings of "classification," and I propose to continue using that name for it. It is, for instance, the first definition under "classification" ("the action of classifying") in the Oxford English Dictionary, where its use is dated from 1790. A possible alternative for "methodological taxonomy" would be "metaclassification," for it involves statements about classification much as metamathematics, for instance, involves statements about mathematics. However, I see no reason why the simpler and more familiar term "taxonomy," which through all its variant usages has always included this subject, should not be defined so as to designate methodological taxonomy.

Taxonomy is the theoretical study of classification, including its bases, principles, procedures, and rules.

I would, then, paraphrase Gregg and say that the subjects of classification are organisms and the subjects of taxonomy are classifications. This usage is etymologically justified, for from its roots (τάξις and νομία) the word "taxonomy" can be freely rendered as "generalization about classification." Most previous definitions of "taxonomy" are not contradicted but only, as regards some of them, made more limiting. The term is not being applied to a different subject, but is only restricted as a result of making distinctions, previously often neglected or misunderstood, among the subjects here separated as systematics, taxonomy, and classification.[5]

FORMS OF CLASSIFICATION: HIERARCHIES AND KEYS

The basic or one might say primitive step in classification is simply the grouping together of individual objects (or concepts) by some system of relationships (or associations) among them. That mental operation plus the designation of each group by a distinctive symbol is the basis of language, and zoological classification at this primitive level is already present in the ordinary vernacular of all languages. The word "cougar," for instance, does not designate any one object (in grammatical terms, it is not a proper noun) but a whole class of similar animals. In this example the class happens to coincide exactly with a zoological species (*Felis concolor*), although as a rule vernacular names are broader than species and rarely coincide with them.[6] Anyone familiar with the animal could readily differentiate it from other wild cats by various diagnostic characters (self-coloring, uniformly

[5] Among others, I previously (e.g., Simpson, 1945) used taxonomy in a broader sense to include both systematics and taxonomy as here defined. There have also been a few really conflicting definitions, for instance by Blackwelder and Boyden (1952), who suggest that "taxonomy can reasonably be restricted to the descriptive, discriminatory, and nomenclatural phase dealing mostly with species and lower categories." I know of no precedents for such usage of "taxonomy" and can see no reason to adopt it. The subject as defined is simply classification at low categorical levels.

[6] The so-called common or vernacular names for species given in field guides and the like are an artificial duplication of the technical nomenclature and are not truly vernacular.

short hair, long tail, rounded ears, large size, etc.), but it is of interest that nontechnical recognition—*identification*—is normally not by
such separate characters but by a mental image of the whole animal.

In some vernaculars and for some classes of objects linguistic classification goes no further than this first, primitive step. For example,
the Kamarokoto Indians have a large vocabulary for particular kinds
of mammals but no generic word equivalent to "mammal" or "animal."
It is, however, at least convenient in everyday speech and is absolutely
necessary in scientific systematics to have a system in which different
levels of generality or inclusion are recognized. There are in particular
two ways in which this is accomplished in practice: first, by overlapping or coincidence of nonidentical classes, and second, by subordination of some classes to others or the inclusion of the former in the
latter. For instance in chemistry there is a class of compounds of
sodium and a class of salts of hydrochloric acid. Neither class is subordinate or included in the other; they have the same categorical rank.
(Precisely what is meant by "category" and by "rank" will be specified
later.) There is a narrower class, that of sodium chloride, that belongs
in both those classes and that is coextensive with the overlap of the
two. Similarly in zoology there can be a class of horned mammals,
a class of two-toed mammals, and a class of horned, two-toed mammals smaller than either and comprising their overlap.

Classification by subordination and inclusion of classes is more usual
in zoology, at least, and more familiar in general. Thus in the vernacular "brown bear" designates a class (corresponding with one or several species, depending on usage and on zoological opinion) that
is subordinate to and included in a larger class designated as "bear."
Other classes such as "polar bear" stand at the same level as "brown
bear" and are also included in the class "bear." This sort of two-word
nomenclature, one designating an inclusive class or set and the two
together designating one of the included classes or subsets, is not
entirely consistent in the vernacular but it is widespread. It has an
evident relationship with technical zoological nomenclature: "bear" =
Ursus; "brown bear" = *Ursus arctos;* "polar bear" = *Ursus maritimus.*
The relationship is not accidental or merely analogous. The technical
system evolved from the vernacular, which originally in Latin (then

the vernacular of learned communication) designated the more inclusive class by a noun and designated each included class by a usually adjectival restricting and defining word or phrase. Linnaeus founded our binomial system by consistently reducing the restricting designation to a single word and by tending (still somewhat inadequately) to separate nomenclature from definition by making the second word only a formal symbol and not also a definition (see Stearn, 1959). That this is only a consistent formalization of an already widespread conceptual scheme or habit of thought is remarkably demonstrated by some close parallels between vernacular and technical binomials. Thus Dennler (1939) has shown that in the South American Indian language Guaraní animals are regularly designated by binomials in which the first word is inclusive (generic) and the second restrictive (specific), exactly as in Linnaean binomials: *tatú*, armadillo; *tatú pará, tatú guasú*, etc., particular kinds of armadillos.

Classification by subordination is of course not restricted to a single step between two levels of classes. In principle the procedure may go on through an indefinitely large number of levels, although in practice, in zoology at least, the number of steps rarely exceeds twenty or twenty-five and may be as small as seven. Such an arrangement is a *hierarchy*. An arrangement by overlap of classes will here be called a *key*, although this has not been the usual definition of that word or, at least, the usual approach to the concept represented by it. Formal definitions adapted to the present approach follow:

A hierarchy is a systematic framework for zoological classification with a sequence of classes (or sets) at different levels in which each class except the lowest includes one or more subordinate classes.

A key is a systematic framework for zoological classification (generally used for identification to the exclusion of other purposes) with a sequence of classes at each level of which more restricted classes are formed by the overlap of two or more classes at the next higher level.

In principle these two frameworks are typically quite distinct, but they may resemble each other in written form and it is sometimes difficult, or even impossible, to be sure whether a given classification is a hierarchy or a key. They are sharply distinguished in principle by the fact that in a hierarchy each class belongs to a single class at the next higher level while in a key each class belongs to two or more classes at the next higher level. Thus:

Hierarchy: Bears
⎧‾‾‾‾‾‾‾‾‾‾‾‾‾‾‾‾‾⎫
Brown bears   Polar bears

Key: Brown animals   Bears   White animals
⎣‾‾‾‾‾‾‾‾‾‾‾‾‾‾⎦   ⎣‾‾‾‾‾‾‾‾‾‾‾‾‾‾⎦
Brown bears   Polar bears

Keys can be diagramed in such a way as clearly to indicate the equivalent level of the more inclusive classes and the way in which their overlap defines the classes at a lower level:

	Bears	*Nonbears*
Brown animals	Brown bears	Classes not here named
White animals	Polar bears	Classes not here named

As keys are written, however, it is often more convenient to take up classes in sequence as if some were subordinate to others even when logically they are at the same level, thus:

1. A. Nonbears
       (not here further classified)
    AA. Bears
       B. Brown . . . . Brown bears
       BB. White . . . . Polar bears

2. A. Brown animals
       B. Bears . . . . Brown bears
       BB. Nonbears . . (not further
                        classified)
    AA. White animals
       B. Bears . . . . Polar bears
       BB. Nonbears . . (not further
                        classified)

These arrangements, especially 1, look like hierarchies. However, each one implies the other and both are just ways of conveying the same information as in the preceding diagram, in which "brown animals," "white animals," "bears," and "nonbears" are all classes at the same level and "brown bears" and "polar bears" are classes at the next lower level related to those higher classes in a key and *not* in a hierarchic way. The most general and significant clue to the difference is that in the key it does not matter which of the classes at the same level—"bears" and "nonbears," "brown animals" and "white animals"—are listed first. We can "key out," that is, identify, an animal just as well starting from the fact that it is white as starting from the fact that it is a bear. Indeed it must often happen that a person recognizing (that is, identifying) a polar bear notices that it is white before he notices the more complex characters that signalize its being a bear.

In a true hierarchy, on the contrary, the sequence of levels and the subordination of some classes to others is an essential element in the system. There is no purely formal or abstract logical reason why the hierarchies

White animals		Brown animals	
Polar bears	Arctic foxes	"Red" foxes	Brown bears

should not be used instead of

Bears		Foxes	
Brown bears	Polar bears	Red foxes	Arctic foxes

—but no zoologist would ever employ the former in preference to the latter. There are excellent reasons, so obvious that they are clearly reflected even in the vernacular, why in a hierarchy being a bear or a fox has priority over being brown or white. The reasons for assignment of priority are generally much more complex and hence on occasion uncertain and disputable, but priority on some basis or other is inherent and necessary in hierarchic classification. The principles and criteria for priority are extraneous to the hierarchy itself; they are *taxonomic* in a broader sense. Their selection and recognition are, indeed, among the most basic problems of taxonomy and are accordingly discussed at considerable length in later chapters.

Because a key involves no principle of priority and has a purely arbitrary conventional sequence, keys are universally considered *artificial*. A hierarchy does involve principles of priority, and to the extent that these principles are derived from real or natural relationships among organisms hierarchic classification is *natural*. (The expression "natural classification" has, however, been used in many other, often conflicting, ways, some of which will be discussed later.) It must be admitted that principles sometimes used in assigning priorities may themselves be more or less arbitrary, in which case a resulting classification may not fully deserve designation as natural. It should also be added that good keys frequently follow a sequence determined by priorities in a hierarchic arrangement of the same animals. In that case, the key may be considered as natural as the hierarchy, but this is derived from the corresponding hierarchic classification and is not inherent in the key as such. In fact, what is actually a hierarchic classification in principle, as defined above, may be *written* as if it were a key; and conversely what is in principle a key may be and not infrequently is written in the form of a hierarchy.

Distinction between a hierarchy and a key and consideration of the relationships between the two are essential to understanding either one of them. Nevertheless, from this point on we will be concerned almost exclusively with classifications in the form of hierarchies. We shall as a rule be using a single form of hierarchy that has been adopted by general agreement for usual zoological classifications and that is the basis for most zoological nomenclature. This hierarchy was developed mainly in the seventeenth and eighteenth centuries and reached nearly definitive form (for zoologists) in the tenth edition, 1758, of the *Systema naturae* by Linnaeus, for which reason it is called the Linnaean hierarchy. Its basic feature is a sequence of seven[7] levels:

Kingdom

Phylum

Class

Order

Family

Genus

Species

The sequence from top to bottom and the customary indentation indicate decreasing scope or inclusiveness of the various levels.

The number of kinds of organisms to be classified has now become so enormous that seven levels are rarely enough in practice. The deficiency has been made up, for the most part, by adding levels designated as *super–* above various of the basic levels and as, successively, *sub–* and *infra–* below them. Numerous proposals to add to the seven basic levels have also been made, but these are not stand-

---

[7] Linnaeus used only five of these. Phylum and family have been added since from the other sources. Linnaeus also used another level, empire, at the top, for the whole world of phenomena.

ardized and are not in general use. It is unnecessary to list all of them, and the usage of any particular taxonomist can be picked up readily enough in his work. Those in widest use are probably *cohort*, between class and order, and *tribe*, usually but not always placed between family and genus. An example of a complete hierarchy used in classification of a large group of animals (mammals; Simpson, 1945) is as follows:

Kingdom
Phylum
Subphylum
Superclass
Class
Subclass
Infraclass
Cohort
Superorder
Order
Suborder
Infraorder
Superfamily
Family
Subfamily
Tribe
Subtribe
Genus
Subgenus
Species
Subspecies

This has twenty-one levels. Use of all possible super-, sub-, and infra- levels between kingdom and subspecies would give thirty-four, probably more than is ever really needed in practice. The use of any particular number of levels is a completely arbitrary convention. It does not correspond with anything in nature but is an artifice imposed by practical necessity in the use of any hierarchy. This does not mean, however, that groups recognized at any given level are necessarily in some sense unreal or do not correspond with naturally definable entities. They may or may not do so, a point to which further reference will be made.

Use of the seven basic Linnaean levels is required by convention, that is, no animal is considered to be satisfactorily classified unless it has been placed implicitly or explicitly in some definite group at each of the seven levels. (In practice, nevertheless, a minority of groups are not completely classified because they have not yet been satisfactorily grouped at some level.) Use of any of the other levels is optional, depending on the taste of each individual classifier and the requirements he finds in the particular group of organisms in question. There is, however, one restriction on freedom of action in this respect: if a subsidiary level is used within any one group it should, as far as possible, be used for all the organisms in that group. For example, if a subfamily is used within a family, then the whole family should be divided into subfamilies and all genera (as far as the data permit) should be placed in one subfamily or another. However, the fact that subfamilies are used in one family of an order does not require that they also be used in other families of the same order.

A few special terms and concepts are here necessary in order to deal clearly with the hierarchy and with hierarchic classifications. In the first place, what actually is classified?—a point so essential and (oddly enough) so frequently misunderstood that the answer is fundamental in everything that is to be said about taxonomy. It seems obvious—rather more obvious than it really is (see especially Chapter 5)—that the real unit in nature, the one thing that is usually completely objective in spite of some marginal cases, is the individual organism. Nevertheless, *an individual never is and cannot be classified.* Classification involves only groups; no entity possible in classification is an individual.[8] An individual may be referred to or placed in a given group. That is often called "classifying" the individual, but that is a misnomer. That process is *identification,* which is not the same

---

[8] It would be mere quibbling to consider the last individual of a group about to become extinct as comprising that group. The first individual of a nascent group (and a valid group arising from one individual must be extremely rare) is likewise not to be considered as constituting the group.

as classification. The existence and nature of a group may also be inferred from one individual, but that, again, is not classification of the individual but is statistical inference from sample to population. What is classified is always a *population,* defined in the broadest sense as any group of organisms systematically related to each other. ("Population" is also sometimes used in systematics in a narrower technical sense, but for clarity we will later distinguish such a population under the term "deme.")

The unit of the Linnaean hierarchy is the *species.* (The status of subspecies and other infraspecific groups of various kinds is discussed in Chapter 5 and need not detain us here.) It is of course composed of a population, a group of organisms defined in certain ways, which are also to be more explicitly discussed later. All other groups at levels above the species are themselves composed of one or more species. These groups, including species, are collectively known as *taxa* and each of them is a *taxon,* defined as follows:

A taxon is a group of real organisms recognized as a formal unit at any level of a hierarchic classification.

Porifera (a phylum), Crustacea (a class), Chelonia (an order), Didelphidae (a family), *Bos* (a genus), and *Homo sapiens* (a species) are all taxa.

We also need precise terms for what have more loosely been called hierarchic levels and for the relationships among them:

A taxonomic category or (in the present context) *simply a category is a class the members of which are all the taxa placed at a given level in a hierarchic classification.*

The rank of a category is either its absolute position in a given hierarchic sequence of categories or its position relative to other categories.

The rank of a taxon is that of the category of which it is a member.

These concepts and some of the others developed in this chapter have recently been especially investigated by Beckner (1959) and Gregg (1954), in the latter case, particularly, by the complex and rigorous use of symbolic logic. (These studies are largely developed from previous, more general work by Woodger, especially 1937, 1948, 1952.) In oversimplified paraphrase of their approach, a taxon (by Gregg called "taxonomic group") is a set the members of which are organisms. Taxonomic names, which are symbols applied to taxa, constitute a name group symbolized by Gregg as N^2. (Individual or proper names, which do not belong to classification, form N^1.) A category (or, as in Gregg, taxonomic category) is a set of sets, the members of which are taxa. Their names (phylum, class, order, etc.) belong to name group N^3. An organism is a member of, or in the vernacular "belongs to," a taxon, in fact to a whole series of them at different ranks. It cannot be a member of (belong to) a category. A man, for instance, cannot logically belong to the category "order," but to the Primates, a taxon which is a member of the category "order" and has the rank of that category.

Beckner symbolizes a taxon of rank j as T_j, with j taking values from 1 to n. T_1 is ordinarily a species, but in some classifications may be a subspecies. More precisely the lowest recognized taxon *is a member of* the lowest recognized category. T_n is ordinarily a kingdom. Each T_j except T_n *is included in* a T_{j+1}. T_1 is the taxon in which no other taxa are included, and T_n is the taxon that includes all other taxa and is not itself included in any other taxon. As noted, in Linnaean classification the number n must be seven or higher and varies from one classification to another, being, for example, twenty-one in Simpson (1945). Beckner further stipulates that "every organism is a member of one and only one taxon of each rank" and also that "there is universal agreement that the number of taxa of rank $j + 1$ should be considerably smaller than the number of rank j." Here, however, actual taxonomic practice is not, and in my opinion should not be, as rigid as this logical scheme. I have already noted that n is not necessarily constant even in different parts of the same classification. For example, subfamilies may be recognized in one family and not in another. For the classification as a whole a rank must be assigned to the category "subfamily" and to the taxa that are members of it, but it is not true that all organisms are members of a taxon of that rank.

Beckner's second generalization here rejected is related to an anomaly developed in some detail by Gregg. Using different words and symbols, Gregg points out that in actual classifications monotypic taxa are quite common, that is, a T_{j+1} commonly includes only one T_j. A rather extreme example is that of the order Tubulidentata (aardvarks), which includes only one living species, so that (excluding a few extinct forms) the order Tubulidentata, family Orycteropodidae, and genus *Orycteropus* are all monotypic. In other words, all have identically the same organisms as members, since all include the single set (species) *Orycteropus afer* and no other sets. Beckner's statement is that by universal agreement this should not occur, but in fact it *does* occur in innumerable classifications that are universally accepted by taxonomists. Gregg accepts that fact, but considers it anomalous or illogical because (in words other than Gregg's) it makes taxa and categories that are logically different identical in practice.

The answer to that objection is that actual classifications involve relationships that are not (up to now) recognized in attempts to apply set theory to hierarchic classification. A T_{j+1} that includes only one T_j does nevertheless differ from that T_j because of a lateral or horizontal relationship or rank equivalence with other T_{j+1}s that do contain more than one T_j. The order Tubulidentata differs from its only included living species *Orycteropus afer* by the very fact of its equation (as to level) with other orders, not species. It symbolizes the fact that *Orycteropus afer* differs from any species of other orders about as much as do two species referred to (included in) any two of the latter orders. That relationship, which will subsequently be developed further, depends on zoological taxonomic considerations not taken into account by Gregg or Beckner.

REMARKS ON SET THEORY AND SYMBOLIC LOGIC

The example just given demonstrates that these particular applications of set theory and symbolic logic to classification do not adequately take into account some relationships that are involved in actual classifications and that do appear to be perfectly logical. That does not imply that these relationships could not be adequately incorporated in some system such as Gregg's or that it would not be worthwhile to do so.

Objection has been made to such systems (also to Woodger's applications of symbolic logic to other aspects of biology) on grounds of tautology and of heuristic failure. The first criticism is that these systems do not say anything that has not already been said more simply (although as a rule less succinctly) in other terms. The second is that they have not (as yet) led to any useful modification of taxonomic practice or to any important taxonomic discoveries. Both criticisms do, in my opinion, have considerable force, but they are not conclusive. In reply to the first, tautological symbolization may nevertheless be useful; algebraic symbols, for instance, are certainly useful and yet say nothing that cannot be expressed at least as clearly in words. Second, the possibilities of such symbolization in application to classification have as yet been little explored and future heuristic value is by no means excluded, and in the meantime demonstration that classification can, if as yet only in part, be related to a rigorous abstract logic does have a certain value.

There are, however, other objections to these systems in their present forms that seem to me to have greater force. As has been demonstrated, they are imperfect and confusing in not taking into account some logical considerations that are essential in actual classification. (There are other omissions besides the one discussed above.) They seem to require or assume the existence of fully separable sets or a rigorous A:not-A (or $x:\bar{x}$) dichotomy that is often inappropriate in dealing with evolving organisms. By implication, at least, they are best or perhaps, at least in part, only applicable to invariable sets definable by characters in common, whereas the members of all taxa are highly variable and in many instances must be defined in other ways. (Beckner's "polytypic concept" has in part met this objection but not fully.) They could more adequately describe typological and more or less artificial classifications such as those usual in the eighteenth century than modern, and specially evolutionary, classifications. (Of course Woodger, Gregg, and Beckner all postulate evolutionary origin of the taxa classified, but they do not seem to me to have been fully successful in getting away from pre-evolutionary typological definitions.)

Some of these and related objections also apply to other approaches to the subject, and it is possible that all could be met by further development of set theory and symbolic logic. At present, however, the inadequacies for my particular purposes do seem to me so serious that I do not propose to follow the subject further or to attempt consistent use of the Woodger–Gregg–Beckner terms and symbols.

The following brief list of some of the most elementary symbols is, however, given by way of example, as being of possible help to those who may care to pursue the subject further (which is recommended), and to demonstrate equivalence with important concepts present throughout this book in different terminology.

ϵ—Membership in a set. $x \epsilon y$ means "x is a member of set y."

$\subset$—Inclusion in a set. $x \subset y$ means "all members of set x are also members of set y."

$=$—Identity. $x = y$ means all members of x are also members of y and all members of y are also members of x, which may also be written "$x \subset y$ and $y \subset x$." [9]

$\neq$—Nonidentity. $x \neq y$ means "some members of x are not members of y."

$\bar{x}$—Complementary. $\bar{x}$ is the set of nonmembers of x.

$\cap$—Overlap. $x \cap y$ is the set of members of both x and y.

$\cup$—Summation. $x \cup y$ is the set of members of either x or y, or both.

THE BASES AND CRITERIA OF CLASSIFICATION

Hierarchic classification, now adequately characterized, inherently involves certain operations and relationships, the formal aspects of which are quite simple although their theoretical bases and their practical application are exceedingly complicated. The primary operation is the (*conceptual*) aggregation of organisms into taxa of lowest rank, T_1s. (In the practice of classification, however, this is not always the *first* operation, and in identification it never is.) Secondary operations consist simply in aggregating those T_1s into taxa of successively higher rank up to T_n, which is in scholastic terms the "highest genus" and in more modern terms the set included in no set. These operations in themselves, regardless of how the grouping is performed, involve two kinds of relationships, or more precisely of relationships between relationships. Inclusion of a T_j in a T_{j+1} involves a relationship of *priority* of the relationship that unites the T_js into the T_{j+1} over the relationship involved in grouping into the T_js themselves. At the same time there is involved a relationship of *equivalence* between the rela-

---

[9] Note the ambiguity in application to taxonomy. By this definition Tubulidentata = *Orycteropus afer*, which is not a valid *taxonomic* statement.

tionship uniting into a particular T_{j+1} and that forming other $T_{j+1}s$ in the same classification.

The relationship of priority is exemplified by the relationship between genus and species and the relationship of equivalence by that between genus and genus within one family (or subfamily or tribe, if those categories are used). In one scholastic usage (not all usages were the same), the terms genus and species were far more generalized than in modern zoology and signalized this relationship in general and not only that between the two particular categories now called genera and species. (See Joseph, 1916; Cain, 1958a.) A species was a group of things similar in essence. A genus was that part of the essence shared by distinct species, hence by extension it was a group of species (at any categorical level) with some attributes in common. *Species* and *genus* were two of the five predicables, which were the most general relations of attributes involved in logical division and predication (assignment to a class). The other predicables were:

Differentia: that part of the essence peculiar to a given species and therefore distinguishing it from other species.

Property: an attribute shared by all members of a species but not part of its essence and not necessary to differentiate it.

Accident: an attribute present in some members of a species but not shared by all and not part of its essence.

The scholastic approach to logical classification is not only totally inadequate but also highly inappropriate for modern taxonomy. It is considered here and again in the next chapter for two reasons. First, it was the background from which zoological classification developed. Its concepts and terminology were familiar to most early taxonomists, including Linnaeus, and strongly influenced them. Second, some present-day taxonomists advocate what is essentially a return to scholastic principles, although some of them, at least, do not seem to be aware that that is what their proposals amount to. They would, for instance, define species by their "essences" ("essential natures," "natures," or similar expression), diagnosing them by "differentia" and grouping them in higher taxa by their "genera" (in the original scholastic sense). "Properties" become nondiagnostic and therefore nonessential but constant characters of species, and "accidents" are variations that do not enter into the "type" (a somewhat later concept) and therefore have no taxonomic significance.

It will, I hope, become increasingly evident in what follows how impractical, indeed how unreal, that system actually is in taxonomy. The scholastic system is, however, one answer, and the answer historically underlying the origin of hierarchic classification, to the question of just how taxa are to be defined and how priorities and equivalences are to be assigned. That question is fundamental, indeed, and much of the rest of this book is devoted to it.

The basis for classification must of course be relationships among organisms, but relationships, in the broad sense, are of innumerable different kinds. It is therefore possible for there to be not only many different classifications but also many different bases for classification. Gilmour (especially 1940, 1951) has particularly emphasized that point, and the following considerations are based in part on his discussion, with considerable modification.

1. A major function of classification is to construct classes (in Neo-Linnaean classification taxa) about which we can make generalizations. (Gilmour considers this *the* primary function and specifies ulterior generalization, but always some generalization enters antecedently into classification, and in some cases no further generalization is sought.)

2. The classes are constructed in connection with a particular purpose, which depends on the kinds of generalizations that are considered pertinent.

3. Some classifications pertain to a wider range of inductions or to more meaningful generalizations than others and are in that sense "better," or more useful. (Gilmour adds that classifications serving a large number of purposes are natural and those serving few purposes are artificial, but this is not the usual nor is it a wholly acceptable usage; "natural" as applied to classifications is ambiguous and disputable, as is discussed elsewhere, but it is hardly ever taken to mean more widely useful. That a natural classification, in some sense of the words, *would be* more meaningful in many respects is a different matter.)

4. "There cannot be one ideal and absolute scheme of classification for any particular set of objects, but . . . must always be a number of classifications, differing in their basis according to the purpose for which they have been constructed" (Gilmour, 1951).

5. Even classifications in the same form (for example, the Linnaean hierarchy), with the same purposes, and based on the same criteria or principles are not unique or uniform. The same data and interpretations may lead to equally "good" or "valid" classifications that nevertheless differ from person to person and from time to time. (This point, also further discussed hereafter, was not made by Gilmour in the papers cited.)

We must thus accept the possibility and in fact the need not only of many classifications but also of many *kinds* of classifications, that is, of classifications based on different sorts of relationships and serving different purposes. For example, two bases for classification that are very ancient and that still serve quite special purposes are *ecological* and *teleological*. Ecological classification defines sets (they should not here be called taxa) according to such criteria as the communities in which the organisms live (for example, swamp plants, soil microorganisms, forest insects) or other environmental factors (for example, alpine plants, fresh-water fishes, cave salamanders). The number of such possibilities is obviously very great and the usefulness of such classifications for particular, rather special purposes is also obvious. Nevertheless no one now proposes the general use of ecological classification especially for the purposes of identification and nomenclature. In modern ecology the organisms involved are identified and named according to *other* classifications *before* they are put into an ecological classification.

Teleological classifications define sets (again, not taxa) according to their usefulness or lack of it, usually with respect to man. Such sets might be, for example: domesticated animals, with meat animals, draft animals, pets, etc. as subsets; edible, nonedible, and poisonous fishes; or herbs classed according to the diseases for which they are considered specifics. Such classifications were formerly widely used, for instance in many of the elaborate early herbals. They are still used in somewhat fragmentary and sometimes rather unsystematic form for limited purposes, for example in ethnobotany, in cook books, in agriculture, or in research on natural remedies. They do not have much general scientific interest, and again in modern usage they require the prior classification of the organisms on some other system.

Although Gilmour is certainly right to insist that one unique and absolute classification is neither necessary nor possible, it is nevertheless true that one *form* of classification must be applied to all organisms (or, for our purposes, at least to all animals), if only for the purpose of nomenclature, whatever other forms of classification may be secondarily adopted for more limited groups and aims. As already indicated, by tacit agreement this general form is that of the Linnaean hierarchy. It is further tacitly agreed that it is most widely meaningful, if not quite necessary, that the basis of such classification should be relationships among the animals themselves, considered as entities in their own right, and not their relationships to other things, such as the environment (ecological classification) or usefulness (teleological).

Even though that much is generally agreed, the actual different kinds of definitions and criteria that may be used continue to be diverse and debatable. The most intuitive approach, and one that is always involved to some extent in practice, is through associations by similarity, and primarily (although now no longer exclusively) by morphological similarity. On the other hand, approach through one kind of association by contiguity, that of evolutionary origin and phylogenetic relationships, is more broadly meaningful for the reasons, among others, that it involves the processes by which it is now known the taxa really originated and that it explains and permits interpretations of the more obvious associations by similarity. There is accordingly a consensus at present that the "best" (most meaningful, most useful for many inductions) method of classification is by evolutionary relationships and not solely on similarity of individuals. There are, however, able and even eminent taxonomists who forcefully reject that conclusion. Moreover, among those who accept it in principle there are differences of opinion as to how practical this approach really is in actual classifications and, to even greater degree, as to just how it is to be applied in practice. The problem is merely stated here in the most general terms. My own opinion, in agreement with an apparent consensus, is that it *is* best and that it *is* practical although, like any system whatever, it involves innumerable subsidiary problems that demand further solution. The rest of this book is for the most part an attempt to expound those solutions.

Bather (1927), who accepted phylogenetic or evolutionary [10] classification as best, even though with some profound misgivings in his later life, noted that:

The biologist demands a system that shall do two things: enable him to refer without fear of error to the particular creatures under discussion; and present the results of his phylogenetic enquiries in ordered form. If the system cannot fulfil both needs, it must at least fulfil one.

The idea that a classification should or could fulfill *only* the purpose of reference is absurd (as Bather of course knew) even though it has been held, more or less *sub rosa*, by some zoologists. "A distinguished zoologist once told the present writer that the search for a natural classification was no part of a museum curator's business. His job was identification, not classification, and he had only to devise some kind of key or card index that would enable one to sort animals quickly and easily into species" (Calman, 1940). The "distinguished zoologist" did not explain how a quick and easy key or index could be devised if it did not involve an orderly expression of some prior enquiries, whether phylogenetic or otherwise, nor, for that matter, did he suggest what scientific purpose would be served by *only* sorting species and nothing else. Certainly a system worth bothering with *must* do *both*: represent an ordering of the results of studies on animals and provide a means of reference to the groups so ordered.

It has already been noted that the means of reference is the assignment of names, now in principle arbitrary, to the taxa of a classification. As an ideal, which has been approached but definitely not achieved in practice, each taxon should have a distinctive name and only one name used by all zoologists. A further ideal, even less nearly achieved as yet, is that its name, once applied, should be permanent. Some lack of permanency is unavoidable and is inherent in the nature of zoological research. With increasing knowledge of animals it is inevitable that

[10] This approach has usually been called "phylogenetic," but I think "evolutionary" is a better label, for reasons that will be made clear later.

two or more species thought to be distinct will be found to represent in fact only one species, or that what was supposed to be one species will be found to be two or more. Increasing knowledge or differences in emphasis will also inevitably require rearrangements, fusions, and separations of taxa above specific rank. In both cases some changes in nomenclature usually ensue.

Personal preferences and emphases also frequently affect rearrangements of supraspecific taxa and may entail changes in nomenclature, although this is not an important cause of nomenclatural change if the taxa are only rearranged, which may of course be done under their old names, and not substantially changed in definition and contents. The fact that no two authorities are likely to adopt precisely the same arrangement of higher taxa also does not necessarily entail confusion of nomenclature as long as all use the same form of classification and system of nomenclature, as most now do.

The most important cause of instability in nomenclature has been the fact that through the past two centuries, with zoologists working all over the world, it has not infrequently happened that more than one name has been independently applied to the same, or nearly the same, taxa. Numerous successive attempts have been made to draw up a code to take care of this and other problems, and this has finally resulted in a single set of International Rules of Zoological Nomenclature, which were again being recodified when this book was written. (There are separate codes for plants and for microorganisms.) A basic rule is that of *priority*, according to which the name of a genus or species [11] is the first name that was validly published in connection with it since 1758. That conventional zero date marked the tenth edition of Linnaeus's *Systema naturae*. (For plants, different dates are used in different groups.)

Offhand it might seem that the principle of priority would suffice and would forthwith stabilize generic and specific names, but in fact it is far from having done so. It frequently happens, even at this late date, that some obscurely published old name turns up with priority over some name that has in the meantime been universally accepted by zoologists. There is no way of ever being certain that the very oldest

[11] Application to families has been urged, but is not legally in effect at this writing.

name has been spotted, and priority has upset the stability of nomenclature almost as much as it has promoted it. Definition of valid publication is complex and is the subject of subsidiary rules that still do not always settle the matter. There are, further, numerous problems as to precisely what currently recognized taxa were really designated by old names.

There has recently been an increasing tendency to meet these and related problems through the action of an International Commission on Zoological Nomenclature, which is empowered to suspend the Rules in particular cases and to designate *nomina conservanda,* names in general use and selected to continue so regardless of priority. The procedure is slow and cumbersome and so far only a minute fraction of zoological names has been stabilized in this way. Moreover some zoologists object to this procedure and continue to insist on strict priority.

The zoological contents of taxa frequently and inevitably change with increases in knowledge and differences of judgment and opinions. That of course constantly raises problems as to whether a current taxon is really the same as one to which a name was originally applied. This problem is met, imperfectly but usually adequately, by the designation of *types.* The type for the name of a species is an individual specimen, and the rule is that regardless of any other contents of the taxon a name belongs to the species in which its type specimen is placed. It frequently happens that types of two or more names are placed in one species, and this is when priority and the lists of *nomina conservanda* are called on to determine which name should actually be used. The type of the name of a genus is the name of a species and is thus indirectly tied to a type specimen. It is generally considered that the type of the name of a family is the name of a genus, but just what is thereby involved has not yet been fully clarified. Proposals have been made to extend the type system (and priority) to names of still higher taxa, above superfamilies, but this provision is not now embodied in the Rules or in general usage. At present the names of those higher taxa, of course much less numerous than names of genera or species, are determined only by consensus and acceptance of authority, and at these levels that informal system seems to work at least as well as the Rules do at lower levels.

Unfortunately the type system has the drawback of having perpetuated an old and now thoroughly unsatisfactory approach to classification. It involves serious confusion that still persists too widely. The very word "type" embodies an idea generally held in the eighteenth and early nineteenth centuries, derived from the scholastics and ultimately from the ancient Greeks, that a species is definable by a fixed pattern (or "essence"), which in another sense is its "type" (also commonly called "archetype"). The type *specimen* then is considered typical in this sense and is used, alone, to define the species and as the unique standard of comparison for identifying other specimens. With recognition that a species is a population, every member of which equally belongs to and exemplifies the species, and that populations always are highly varied, it would seem obvious that a species simply cannot have a type in that sense or one useful for those purposes. A nomenclatural type is simply something to which a name is attached by purely legalistic convention. It should have nothing to do with the nonnomenclatural processes of defining the species and should have no special role in identifying other specimens. Modern taxonomists are becoming increasingly careful in making this distinction, but the old confusion still permeates much of zoological thought and procedure. It is, indeed, perpetuated by the Rules, which continue to speak of the types of species, genera, and so on, when they should refer only to the types of *names*. It is nominalistic absurdity to confuse a set of objects with the name or symbol for that set.

It is of course now widely recognized that a taxon cannot be adequately characterized from its type alone.[12] Clearly for nomenclatural purposes the name of a species should have a single type, but for purposes of typifying the species or defining it other specimens are necessary (or at least their existence must be assumed). Many taxonomists who recognize these facts nevertheless still confuse the type of a name with the typification or other representation of the group named. They have therefore proposed that many other kinds of "types" should be recognized. That has been carried to truly ridiculous extremes and

[12] In some instances only one specimen of a species is actually known, and then necessarily the name-bearer is the same as the sample (the *hypodigm* as defined elsewhere) on which definition is based. Nevertheless the functions of name-bearer, nomenclatural type, and of taxonomic sample are quite distinct, and the definition must assume the existence of other and variant individuals in the species in nature.

literally dozens of terms for kinds of "types" have been proposed: topotype, homotype, metatype, and so on. For purposes of nomenclature, which are the *only* purposes that types can really serve in modern taxonomy, only four terms are really needed:

Type (or *holotype*): the single type of a name.

Syntypes: two or more specimens all of which were treated as types in the original proposal of a name—a procedure now considered inacceptable but formerly widely used.

Lectotype: one of two or more syntypes subsequently selected to play the role of the single type.

Neotype: a substitute for a type that has been lost or (in some usages) is otherwise inadequate for the type role. (Neotypes have not hitherto been recognized in the Rules, but there are proposals to incorporate them therein.)

One other term should be defined because it is so commonly used even in current taxonomic literature, although (in my opinion) it is both unnecessary and confusing:

Paratype: a specimen other than the type but designated by the original author of a name as having been used by him in defining a species.

By convention, technical names of animals are to be treated as if they were Neo-Latin. Many are in fact Latin: *Homo sapiens, Bos taurus.* Many, especially names of genera, are Greek transcribed into Latin: *Synoplotherium, Xenochoerus.* They may, however, be of nonclassical origin or even frankly invented: *Lama* (Peruvian Indian), *Dugong* (Malay), *Kogia* (a bad English pun for "codger"), *Tatera* (invented).

The name of a genus is a single word, capitalized and treated as a Latin noun in the nominative singular. The name of a species consists of two words, which is the reason why Linnaean nomenclature is called *binomial* or *binary.* The first word is the name of the genus to which the species belongs. The second word, called simply the specific or sometimes the trivial name, is not capitalized and is treated either as a Latin noun in apposition, a Latin noun (especially a proper name) in the genitive, or a Latin adjective, which must agree with the generic name in number (singular) and in gender. Names of genera and species are customarily and preferably italicized. Names of subgenera have the same form as those of genera and are written in parentheses after the name of the genus to which they belong. Names of subspecies consist of three words, the first two the name of the species and the third a word treated like the trivial name of the species. By convention, if a species is subdivided into subspecies, one of the subspecies must have the same trivial epithet as the species: One of the many subspecies of *Rattus rattus* is *Rattus rattus rattus*. (The identity of generic and specific names is, however, not required and is unusual in this example.)

Names of taxa above the genus (or supergenus, which is almost never used) are single words treated as Latin nouns in the plural. They are capitalized but not italicized. Through superfamilies, they are customarily formed by adding conventional endings to the stem of the name of one of the included genera. The genus of which the name is so used is generally considered the type of the higher taxon.

Superfamily: –oidea (for vertebrates) or –aceae (for invertebrates).
   Example: Hominoidea (from the stem *homin–* of *Homo*).
Family: –idae. *Example:* Hominidae.
Subfamily: –inae. *Example:* Homininae.

Names of taxa above superfamilies are, at present, usually not derived from generic names, although they may be, and usually do not have uniform endings. Provision of uniform endings for all taxa or particularly for orders has often been proposed and inclusion of some such provision in the Rules is under consideration, but this is not now general practice and in my personal opinion it would be undesirable for the animal kingdom as a whole. In a few special groups this practice has, indeed, become established. For example, the names of orders of birds now most commonly used are compounds of the stem of a generic name and the termination –iformes.

In early classifications the names of taxa were usually descriptive if not definitive. The Carnivora, for instance, were so named because they were believed all to be carnivorous, even though it was recognized that not all carnivorous animals are Carnivora. Now we know that some Carnivora are, in fact, strictly herbivorous. It has already been mentioned that specific nomenclature developed from what were originally brief Latin *definitions* (or the scholastic *differentia*) of the species. That history has been an element in the slow and still not entirely complete recognition of the fact that nomenclature and the recognition and definition of the taxa named are two quite separate and distinct things. With the reduction of the second part of specific names to single words, the enormously increasing number of known species, and realization that many higher taxa must, like the Carnivora, include exceptions to any one-word descriptive designation, it became evident that practically no name can constitute a definition or diagnosis and that not all can be correctly descriptive. Names that are incidentally descriptive have of course been inherited from the past and are still frequently being coined. Nevertheless it is now recognized that a technical zoological name is not a characterization of a taxon but is a purely conventional label. What it *means* is the taxon and not something suggested by its etymological origins, just as George Fletcher means a particular man and not "a farmer who makes arrows." *Basilosaurus* does not mean "king of the reptiles" but a genus of extinct marine mammals.

Working classifiers must, of course, be familiar with all the innumerable details of the zoological nomenclatural system. For that purpose Mayr, Linsley, and Usinger (1953) have provided an excellent introduction, and the Rules and numerous other publications of the International Commission must also be studied. The summary and elementary discussion here suffices for present purposes: to supply essential background and suggest the relationship between nomenclature and the other aspects of taxonomy with which the rest of this book is concerned.

Zoological nomenclature is merely a labeling of the taxa of classifications. It has no other function in taxonomy. It provides a vocabulary for writing and talking about animals, and so is absolutely essential to zoology, but it has no other zoological or scientific interest in itself. It is an arbitrary device that has become an enormously complex, strictly formal, rigidly legalistic system. Some zoologists do seem to enjoy those legal, essentially nonzoological, seemingly endless rules, discussions, and operations, but for most they are a necessary evil taking begrudged time from more important matters. For all its progressive modifications and esoteric intricacies, zoological nomenclature is still far from solving all its problems, or from solving any one of them to the satisfaction of all. Nevertheless it does somehow muddle along well enough to be workable. We shall not henceforth be directly concerned with it, even though we will of course be using its vocabulary.

2

The Development of Modern Taxonomy

Taxonomy has a long history, going back to the ancient Greeks and to forerunners even less sophisticated in systematics. Our interest here is centered on modern taxonomy itself, and we shall largely ignore developments before about the middle of the eighteenth century, when taxonomy as a distinct branch of science had already definitely taken form. Even for the last two hundred years no sequential history is here attempted.[1] The concern here is, rather, with a few of the most important principles, concepts, and systems especially involved in the historical origins of modern taxonomy. Among these are: the theoretical criteria for classification; ideas of natural classification; the effects of the discovery of evolution; and some formal aspects of various visualizations or patterns of relationships among organisms. (The formal framework of hierarchic classification was sufficiently discussed in the first chapter, but patterns of relationships are a different matter.) Finally, an attempt will be made to see just what distinguishes "the new systematics" from the old and to characterize modern taxonomy, which has progressed beyond the original "new systematics" of some twenty years or more ago.

[1] To my knowledge, no adequate treatment or even sufficient summary of the history of taxonomy or of classification has been published. There are a few useful books bearing on the subject. Sprague et al. (1950) have compiled a very brief history. Zimmermann (1953) has given many annotated examples of pre-Darwinian taxonomy (post-Darwinian examples are inadequately and poorly selected), and Gregory (1910) has followed the classification of a single group (mammals) historically in some detail. There are, of course, also scores or hundreds of shorter studies on particular aspects, of which Cain (1958) is just one outstanding recent example.

SCHOLASTICISM AND LINNAEUS

As mentioned in Chapter 1, among all the kinds of relationships (or associations) that can enter into classification the two of primary interest here are those of similarities among organisms and those involved in their evolution and descent. The two are by no means unrelated or sharply separable. In modern taxonomy—with some dissent, to be duly noted—the *concepts* and *definitions* of categories and taxa are reached mainly by evolutionary considerations, but the evidence that validates the concepts and demonstrates that the definitions have been met is derived in considerable part from data as to similarity. Certainly neither one can be omitted or even given precedence over the other. They are simply different parts of the same process, of understanding the order among organisms in nature and of making classifications on that basis.

That relationship could not begin to be understood until the truth of evolution was recognized. Many had inklings about evolution before him, but Lamarck was the first to introduce the concept into taxonomy with any considerable clarity. Even for Lamarck, evolution had very little impact on the practice of classification, and only with Darwin did a slow and still incomplete movement toward a fully evolutionary taxonomy begin. Yet the form of classification still in use was established before Lamarck and so were many taxonomic principles, some still accepted and some others with enduring influence even though rejected in their original form, at least.

In order to understand the situation at the time of Linnaeus, it is necessary to go back again to scholastic logic, Aristotelianism, and Thomism.[2] That body of thought was particularly fundamental to Linnaeus and permeates his work. According to that system of logic, introduced briefly in the first chapter, the whole "essence" of a "species" (in the nonmodern scholastic sense) consists of its "genus" (same sense) plus its "differentia." However, Cain (1958a), to whose careful analysis I am greatly indebted throughout this section, points

[2] If I may be permitted a personal remark, I am somewhat reluctant to do this. I tend to agree with Roger Bacon that the study of Aristotle increases ignorance. Nevertheless, the founders of taxonomy were themselves students of Aristotle and Aquinas (among many others of that lineage) so that the subject is to some extent necessary for my purpose.

out that this is strictly true only if there is a logical relationship between the genus and its species and among its various species. They must use the same *fundamentum divisionis*. The classical example is the differentiation of the genus "triangle" into the species "isosceles," "equilateral," and "scalene," which have the same *fundamentum divisionis* and, as they should, completely exhaust the logical possibilities under triangle. The genus "triangle" may also be divided into "right-angled," "obtuse-angled," "acute-angled" by exhaustive use of another *fundamentum divisionis*. But a single system or classification cannot logically use both *fundamenta*. In a more biological example, a genus "legged animal" may have species "two-legged," "four-legged," "six-legged," etc., or species "short-legged" and "long-legged," but it cannot simultaneously have both. (These considerations are fully discussed in Joseph, 1916.)

A rigidly logical procedure of that sort produces a *taxonomy of analyzed entities*. Attempts to apply it to the classification of organisms showed to Linnaeus and his predecessors and contemporaries that it is inadequate and in part inappropriate for that purpose, and that in animal (or plant) taxonomy only a *taxonomy of unanalyzed entities* is practically possible. One major difficulty, still far from being met by recent taxonomists who would classify by "essential natures," is the impossibility in most cases of distinguishing "essence" from "property" (an attribute shared by all members of a species but not part of its genus or differentia) or even from "accident" (an incidental attribute of some members of a species). Another is that in logic the same individual within any *one* genus may belong to any number of *distinct* species according to the various different *fundamenta divisionis* applied, and that this is completely inadmissible in a zoological classification of genera and species. It was in practice further found to be difficult, in some groups quite impossible, to use the same *fundamentum* for all zoological species in one zoological genus or to keep the "differentia" related to the logical genus. For example, an authoritative eighteenth-century classification of mammals (Pennant, 1781) includes the "genera" (in the old broad sense; these taxa have higher rank in modern usage) "whole-hoofed" and "digitated," which are substantially logical in having the same *fundamentum* when considered as "species" of a higher "genus." But under "digitated" are included the "species" (also old broad sense) "anthropomorphous frugivorous," "with large canine teeth," and "without canine teeth." The last two have the same *fundamentum divisionis,* but the first has a completely different one, and none has a *fundamentum* logically related to the genus.

In retrospect the scholastic approach has several other faults (as a basis for zoological classification) so serious as to make it completely inacceptable to the majority of modern taxonomists. These drawbacks could not very well be and were not fully recognized in Linnaeus's day, and even now are not universally recognized. Among the most serious of them are the following two. First, what were called "properties" and what were called "accidents," both excluded from the "essence" and therefore from the definition, now appear to be essentially characteristic of a taxon, necessary in its proper description and commonly also in its practical definition. That is almost always true of "properties," which are as a rule genetical in basis. It is frequently true also of "accidents," variations within a species, which are always as important for understanding the species and often as essential for definition and identification as are characters constant in the species. Second, even as a taxonomy of unanalyzed entities application of scholastic logic produced a classification of *characters,* not of *organisms.* A classification of organisms was also produced, but this was parallel and really external to the logical system, which defined attributes of organisms and not organisms themselves. Linnaeus and some others did dimly see this difficulty but did not solve the resulting problem.

Part of the eighteenth-century solution to the problems then recognized was the simultaneous use of more than one *fundamentum divisionis* in the "differentia." A famous classical example is the definition of man as an unfeathered biped. Neither "unfeathered" nor "biped" can be considered the "genus" and the other the "species." In our terminology, they are both associations by similarity that define sets of the same rank, and "man" is a subset defined by the *overlap* of the other two. (As it happens, man is not really the only member of that subset, as was intended.) In short, this is in a somewhat cryptic way a key classification, as that term is defined here. This device is almost universal in Linnaeus; one example is in the definition (logically the "differentia") of the bird *Fulica atra:* "fronte calva, corpore nigro, digitis lobatus," with three obviously different *fundamenta divisionis.* Nevertheless the use of one *fundamentum* at any one level and of logically related *fundamenta* at different levels was usually considered desirable as an ideal, even though it plainly could not be achieved in practice for any considerable group of animals, still less for the whole animal kingdom. Thus Klein (1751) managed a sequence of "quadrupeds," under which are "unguiculates" and "ungulates," and under the latter "one-hoofed," "two-hoofed," "three-hoofed," "four-hoofed," and "five-hoofed," but other *fundamenta* had to be brought in for genera and species (in the modern sense). At about the same time Brisson (1762) subdivided all the ungulates on (with slight exceptions) a single *fundamentum* relating to the teeth, but required other *fundamenta* at both higher and lower levels. That incidentally brings up the question of how the *fundamentum* is to be chosen: why for Klein it should relate to toes and for Brisson at essentially the same level to teeth.

Before moving on to other aspects of the eighteenth-century taxonomy, it may be pointed out parenthetically that it is now generally considered desirable to follow both of the procedures discussed in the last paragraph, even though in modified form, in different terms, and usually for different reasons. The use of multiple characters at each level in classification is universally recommended, and the use of the same kinds of characters (same sets of *fundamenta divisionis*) at any one level, although not universal, is preferred when practical.

Here we return again, as we must repeatedly in successive contexts, to the problem of ranking or priority of characters used in hierarchic classification. If the *fundamenta divisionis* of "species" (scholastic usage) are logically related to those of "genera" (same), the problem tends to solve itself, for the "differentia" are then included in and are exhaustive subdivisions of the "genus." "Triangle" obviously belongs at the next higher rank to "isosceles," "equilateral," and "scalene," and "legged" above "two-legged," "four-legged," etc. When, however, the *fundamenta* differ from level to level, ranking must depend on some considerations outside of abstract logic. A distinction is usually made between *a priori, theoretical,* or *deductive* and *a posteriori,* or *empirical,* or *inductive* approaches. The former involves thinking out beforehand what kinds of attributes should be more important or least dispensable in the organisms to be classified and then deducing a theoretical sequence of ranks on that basis. The latter involves first observing the actual distribution of attributes in the organisms and then deciding on that basis how to assign priorities, the usual but not necessary criterion being extent of constancy among members of the group.

The difference between these approaches is frequently obscure, at least to me, in particular instances, and I question whether this is a clear-cut and exhaustive dichotomy. In spite of the fascinating and lucid expositions by Cain (especially 1958 and 1959b; see also, among others, Stearn, 1959), I am uncertain as to which system Linnaeus really intended to follow in some instances, and it does seem that his practice was a complex mixture of the two. Evidently he considered the a priori method as theoretically more desirable and related this to the concept of a natural system (referred to again later), but found some degree of empiricism unavoidable. His aphorism that the characters do not make the genus but the genus makes the characters seems to imply that species were first assembled into a genus and that the characters of the genus were then empirically determined from those of the assemblage. However, the words may have carried other connotations in the quite different logic and philosophy of his day.

In any event, one can agree with Cain and others that one aspect of the general movement in taxonomy from Linnaeus to Darwin was some degree of lessening apriorism and increasing empiricism. Personally, I no longer fully agree with Cain when he suggests that desirable progress in empiricism was prematurely checked by Darwin's influence and that one apriorism was then substituted for another. I disagree (mildly) on several grounds. First, I suspect that nineteenth-century empiricism had little more to offer when it was checked or perhaps rather redirected by Darwin. Second, the evolutionary system that has developed from but far beyond Darwin seems to me truly neither a priori nor empirical but rather an elegant union of the two, among other things. Third, the situation was actually far more complex than a simple bout of apriorism versus empiricism, and in other, perhaps more important respects the Darwinian revolution was overdue. It may be added that a school of more or less pure empiricism by no means disappeared after Darwin and has vociferous advocates today, but I know of no modern taxonomists who are strict apriorists.

SOME OTHER PRE-DARWINIAN APPROACHES

Adanson (1727–1806), a young contemporary and rival of Linnaeus (1707–1778), is generally cited as a founder of empiricism in taxonomy. (See among many others Bather, 1927, and Cain, 1959a.) He happened to be a botanist, but his principles are equally applicable in zoology. His taxonomy (which is not necessarily to say his classification) is in important respects more acceptable today than that of Linnaeus or, say, Cuvier. It avoided the obfuscations of deductive logical systems and the erroneous bases of theories now discarded. It was restricted as far as possible, which was and is not always very far, to simple evaluation of observed data. On the other hand, the very fact that it was to a great extent nontheoretical meant that it lacked an explanatory element, a meaningfulness, a judgment that can only be provided in taxonomy and in classification by some form of theoretical evaluation of the data. Adanson's approach was, however, the one best suited to provide the evidence and basis for evaluations and criteria that were later to be supplied by evolutionary theory. Linnaeus, Cuvier, and Owen, and their successors, among many others of earlier theoretical schools, built into taxonomy a great deal of lumber that has had to be cleared away laboriously. Adanson and his followers built a generally sound foundation. One cannot say of modern Adansonians (who do not call themselves that) that they are wrong, but only that their work is shallow and incomplete.

The leading principle of this school is quite simple: observe and record as many characteristics as possible and then group them according to a majority of shared characters. A species consists of individuals with a maximum number of shared characters. A genus consists of species with a maximum of shared characters. And so on. The necessary system of precedence of characters and ranks of taxa [3] is thus built up step by step. Moreover, Adanson (for example, 1763), Vicq d'Azyr (for example, 1792), and others of their general school ruled that membership in a group did not require sharing *all* its characters and did not require expulsion of exceptions to definitions or con-

---

[3] I use this and some other terms for convenience and brevity in referring to work by taxonomists who never heard the terms. The term *taxa* is recent, but Adanson's classifications contained taxa, and calling them that should rather clarify than falsify his thought.

sequent demotion of the group to lower rank. They thus avoided some, but not all, of the now objectionable features of typology (see below) and of scholastic logic in taxonomy. It sufficed to form a taxon if each member had a *majority* of the *total* attributes of a taxon. By this principle, using letters to represent attributes and columns to represent individuals, the following would be two valid species:

Individuals:	1 2 3		4 5 6
	a a b		a a b
Attributes	b b c		e e f
	c d d		f g g
	Species I		Species II

The "species" may be *defined* as follows:

Species I is a group of individuals among which the attributes *a, b, c, d* occur and each of which has a majority (that is, three) of those attributes.

Species II is a group of individuals among which the attributes *a, b, e, f, g* occur and each of which has a majority (that is, three) of those attributes.

The "species" may be *diagnosed* (that is, distinguished from each other rather than delimited each in terms of its own characters) as follows:

In Species I each member has at least one of the attributes *c, d,* neither of which occurs in Species II, and in Species II each member has at least one of the attributes *e, f, g,* none of which occurs in Species I.

This is a perfectly definite, taxonomically valid and meaningful procedure that involves *no* characters in common, despite the fact that even now some taxonomists consider characters in common essential to concepts of taxa.

The defining attributes do not appear in all individuals of either species; each individual in the first species has only one and those in the second have none of their three attributes common to the whole species; and all individuals of both species have attributes that enter into the definition of the other species. The example is oversimple and is also more extreme than early taxonomists of this school would probably have accepted, but it demonstrates the principle. The principle (which he calls "polytypic") has been elucidated at greater length and in ultramodern terms by Beckner (1959). For future reference, note that Species II has no type in the typological sense of a pattern common to all its species, and that the type, in that sense, of Species I consists only of *b*, which is not exclusive to Species I but also occurs in Species II. Note also that Species II has no "differentia" and that the two, although to us they seem obviously related, have no scholastic or Linnaean "genus."

The principle in this form still lacks the interpretation and evaluation to be supplied by evolutionary taxonomy, but it is highly pertinent to that taxonomy and is frequently exemplified in modern classifications. On the other hand, it was not and is not invariably accepted by all empiricists, still less by all nonevolutionary taxonomists.

Cuvier (1769–1832) perhaps had more influence on taxonomy than anyone else between Linnaeus and Darwin—his life spanned the whole interval between theirs. He is especially remembered as the founder of the science of paleontology, which is a mild exaggeration, and the propounder of the principle of organic integration or correlation. Thompson (1952) has written as follows of that principle and its relationship to "systematics" ($=$ taxonomy, as here defined):

Cuvier defined, with Gallic clarity, what he saw as the basic philosophical [4] principle of rational anatomy and, therefore, of rational systematics. "Every organized being," he said, "forms a *whole*, an unique and closed system, whose parts mutually correspond and contribute to the same ultimate action by reciprocal reaction. None of these parts can change unless the others also change, and, consequently, every one of them taken separately indicates and implies all the others." . . . He was convinced that the form is conditioned by function; and that the relation between form and function is governed by laws as inevitable as the laws of metaphysics and mathematics. Interpreted in this way, the specific type acquires the coherence and necessity of a geometrical definition from which we can deduce all the attributes of the object of the definition.

This passage does correctly summarize one aspect of Cuvier's theoretical system, but it is fallacious in the way it relates this to taxonomy. The fact that organisms are integrated organismic wholes is the pro-

[4] I would say "theoretical." G. G. S.

claimed basis for a whole school of modern systematics (see discussion in Beckner, 1959, and his references), but it would certainly have been no news to any taxonomist long before the time of Cuvier. The conclusion now drawn from this apparently rather banal fact is that one should classify whole organisms, not their parts or their attributes. But the Cuvierian principle does not lead to that conclusion and indeed even militates against it. If any one part implied the whole and if all attributes could be deduced from a "type," then a classification of parts and of "types," rather than whole and real organisms, would not only suffice but also be theoretically preferable. The condition is contrary to fact, because a part does not adequately imply the whole even in an integrated organism,[5] and the relationship of "types" to attributes in practice was more inductive than deductive.

Even apart from its inadequacy in its own field, the Cuvierian principle does not in fact provide any workable basis for classification. Neither Cuvier himself nor his many nineteenth-century followers nor the few twentieth-century Neo-Cuvierians, like Thompson, has ever really produced a classification with that theoretical foundation. Thompson discards all the other currently used bases for classification one by one, and earlier in the same paper (1952, p. 5) makes this significant statement:

The good systematist develops what the medieval philosophers called a *habitus*, which is more than a habit and is better designated by its other name of *secunda natura*. Perhaps, like a tennis player or a musician, he works best when he does not get too introspective about what he is doing.

Up to a point, any experienced taxonomist will agree. He might add, however, (as of course Thompson knows, and himself exemplifies) that, in terms of the same analogy, the musician only reaches this point after he has learned conventions and methods so thoroughly that they have *become* second nature, and further that the composition of the music required the development and application of very definite and conscious theories and principles.

The Cuvierian taxonomy (notably in Cuvier, 1835), which like the

[5] The falsity of the Cuvierian principle was early demonstrated by the famous case of the chalicotheres, whose feet imply a carnivore but whose skulls imply an ungulate. The principle survives in folklore as the belief that paleontologists can reconstruct a whole skeleton from a single bone, which all paleontologists know, to their sorrow, to be untrue.

Linnaean has recently been skillfully analyzed by Cain (1959b), was mainly based on principles distinct from that of organic correlation. Cuvier used the hierarchy; this was the "method." He perceived that to make this method meaningful there must be as basis for ranking some defining attributes higher than others. In general, he attempted to establish ranks of defining attributes and so of defined taxa in a deductive or a priori way. He argued, for instance, that sensation and motion are the attributes of highest rank or most influential in animals, because it is these that (in his view) *make* them animals. Nevertheless he also frequently argued for priority on the basis of constancy, that is of occurrence in a larger number of groups of organisms. For example, the circulatory system is "less essential" than the digestive system because not all animals with a digestive system also have a circulatory system. This seems to be a sort of crypto-empiricism, at least.

As Cain has emphasized, this double approach led Cuvier into a vicious circle that tended to make his whole system fallacious. How can one determine the constancy of characters in a group, hence their rank in a hierarchy of definitions, unless the group has already been formed on some other basis? The question is by no means confined to Cuvierian taxonomy but arises over and over again in the work of many schools of thought from the earliest classifications down to the present day. It is all very well to say with Linnaeus that the genus gives the characters. What, then, gives the genus? In effect, Cuvier said that the characters do, which closes the circle, and there is, in principle, no way to break in and start the process to working. Linnaeus's answer was that God gave the genus, but that is not a very practical means of recognizing it. The deductive part of Cuvier's system was another attempt to bring in criteria from outside the vicious circle, but in many instances if not in all, Cuvier's deductions seem merely to be a circling back through prior inductions. Indeed, as Cain has also remarked in other words, many publicly deductive systems were privately built up inductively before being published.

There seem to be only two valid and practical ways to avoid this impasse. One is to maintain a strictly inductive, entirely empirical approach, but that is shallow, as already remarked, and also involves complex and sometimes insuperable difficulties of ranking, as will be further discussed. The other is to bring in criteria for ranking, notably evolutionary deductions, that, although *derived* inductively are independent of the apriorism of Cuvier and many others and external to the vicious circle.

One other aspect of Cuvier's system requires special notice, as it is also representative of a large body of pre-Darwinian (and indeed also a significant but smaller body of post-Darwinian) thought. Having in theory established the ranks of attributes, Cuvier did bring in his principle of organismic correlation as indicating that all parts of the body would then be consistent with the same classification which can therefore be based on a general *plan* of construction. In accordance with this view, he discarded the very ancient idea of a single ladder of nature or chain of being, supported (with modifications) by Lamarck in Cuvier's day and by many others before and a few since. Instead of a gradual sequence, he recognized a hierarchy of multiple, discrete "plans." Among animals he recognized four most general "plans," defining four high-level taxa, called "embranchements" and named Vertebrata, Mollusca, Articulata, and Radiata. Cuvier's "plans" are the "types" of taxonomic typology, a subject to which we must now turn.

TYPOLOGY

Typology stems from Plato and his sources and came into taxonomy along with Aristotelian, Neo-Platonic, scholastic, and Thomist philosophy and logic. As a distinct school of taxonomy it can be fairly well defined in the *Naturphilosophen* and the sequence of Goethe (1749–1832), Cuvier (1769–1832), Oken (1779–1851), and Owen (1804–1892). (For quotations and summaries of the views of most of those involved since ancient times, see Zimmermann, 1953.) Its basic concept has, however, been far more pervasive than the particular systems of the taxonomists named as the leading theoretical typologists. It is not only a particular school of taxonomy but also an approach or conceptualization that is not alternative to but has been an element in almost all schools at one time or another, even the empirical and the evolutionary schools which in different ways and among various practitioners have done the most to break away from typology.

The basic concept of typology is this: every natural group of organisms, hence every natural taxon in classification, has an invariant, generalized or idealized pattern shared by all members of the group. The pattern of a lower taxon is superimposed upon that of a higher taxon to which it belongs, without essentially modifying the higher pattern. Lower patterns include variations on the theme of the higher pattern and they fill in details, different for different taxa at the same level, within the more generalized, less detailed higher pattern. The most detailed pattern is that of a species. (The question of infraspecific taxa need not be considered at this point.) Variations within a species, the "accidents" of the scholastics, are a nuisance but (or because) they have no taxonomic significance. Numerous different terms have been given to these idealized patterns, often simply "type" but also "archetype," "Bauplan" or "structural plan," "Morphotypus" or "morphotype," [6] "plan" and others.

Now we can see more clearly how there arose the previously mentioned confusion when the term and concept "type" came to be used also in technical nomenclature. On typological principles any specimen of a species embodies its "type," and hence is adequate for defining the species and as a standard of comparison. No conflict and indeed usually no difference was perceived between using the same single specimen as a morphological type and a nomenclatural type. Additional specimens, admittedly unnecessary for nomenclatural purposes, were considered taxonomically useful only in helping to distinguish the type from accidental individual variations. Even if the typological premises were granted, later taxonomists frequently felt that there was some kind of difference in "type" between, say, male and female or larva and imago, although in principle both had the same specific type. Hence the proposal of "paratypes," a compromise entirely unjustified on either side although still in common use. Nomenclature requires and indeed can tolerate no type but *the* type or holotype. If variations within a species do not have any taxonomic significance, then again additions to the specific type are beside the point. If they do have taxonomic significance, as virtually all modern

---

[6] A distinction is sometimes made between "Bauplan" and "Morphotypus," the Bauplan being a broader, more abstracted pattern and the Morphotypus a more specific or detailed variant pattern in one of various groups with the same Bauplan (see Zangerl, 1948). Both are types in the general typological sense, and the distinction, which in any case seems rather obscure, need not concern us here.

taxonomists would agree, then *all* known variations should be taken into account, not just those that happen to be shown by selected "paratypes." Henceforth in order to avoid that sort of confusion I shall use Owen's term *"archetype"* for a type in the typological sense and shall use "type" to refer only to nomenclatural types.[7]

The scholastic approach, as exemplified in Linnaeus, leads directly to the archetype concept because the "essence" must be common to all members of a species. The use of multiple *fundamenta divisionis* thus inevitably led to the concept of a whole pattern in common, and hence to envisioning an archetype. Even the recognition that in classification, as opposed to abstract logic, variations enter the picture did not get away from this. Thus Joseph (1916) considers classification as a substitute for and not an application of logical definition, but says of classification that it "attempts to establish types; it selects some particular characteristics as determining the type of any species. . . . The description of the type will serve for [that is, in place of] definition."

Early empiricism was a step away from typological theory and in part a reaction against it. Previously in this chapter I illustrated the fact that empirical principles, carried to a logical conclusion, can construct and define taxa for which no archetype can be abstracted. Nevertheless the trend of empiricism has been in the direction of typological practice if not theory. For example, two important recent proposals for the empirical determination of "relationships" or "affinity" (Sokal and Michener, 1958, and Cain and Harrison, 1958) both measure conformance to patterns of shared characters and are quantifications of essentially typological procedures even though these authors would subsequently *interpret* their results in nontypological ways.

Evolutionary classification, although now often contrasted with typological, has in fact often been pervaded by typological concepts. Darwin's taxonomy was still basically typological, and the word "type," frequent in his works, can often be read "archetype" without distorting the thought. More naïve evolutionary taxonomists have sometimes

[7] It would certainly be better to escape the aura of "typical" altogether and to give another name to nomenclatural types. I once (Simpson, 1940) suggested that some such term as "onomatophore" would be appropriate, but I did not and still do not adopt my own suggestion and no one else has done so, simply because the dead hand of the past weighs too heavily on us all.

equated "archetype" with "common ancestor," although Darwin (1859) clearly pointed out in other terms that this is false. It is still quite widely assumed that the number of characters in common is directly proportional to closeness of phylogenetic affinity, which is a half-truth, perhaps even a three-quarters truth, but is open to so much qualification and so many exceptions that it is not a *general* principle of truly evolutionary classification. Kiriakoff (1959) goes too far when he says that "most of the biologists who consider themselves to be phylogeneticists are typologists," but there is certainly some truth in the accusation.

So much for the spread of typology into all schools of taxonomy, including some that usually oppose typology in principle. Avowed typologists (under that or other names), as exemplified by Goethe or Owen, had a philosophical-theoretical background rejected by most later taxonomists even when they were typological in practice. Their belief was that the archetype is in some way the reality and that organisms are merely the shadows, reflections, or imperfect and evanescent embodiments of that transcendental reality. Archetypes were considered philosophically as Platonic *ideas,* or theologically as patterns of divine creation. Whatever else one may think about them, such metaphysical beliefs have no place in science and they have no heuristic value.

Along with typological taxonomy, based on the same principles, and forming part of the same school in systematics, was *idealistic morphology,* the aim of which was to construct (or recognize) the archetypes on which typological classifications were based. That school has not died out but on the contrary has had a recent renaissance at the hands of Naef, Kälin, and other European comparative anatomists (summary and citations in Zangerl, 1948). The claim is made that the archetype (Bauplan, Morphotypus) is at least a useful tool, at most an expression of anatomical relationships really existing in nature, and that no implications either metaphysical or directly phylogenetic are involved. I must, however, agree with Block (1956) that (partly unconscious) contamination by metaphysics has not been eliminated.

The observation and interpretation of characters in common in groups of organisms must of course always be one of the procedures in any system of taxonomy. Beyond that rather obvious point, however, I maintain that typological theory or the strictly typological approach should have no part in modern taxonomy. The objections should be evident to anyone who reads this whole book, but a few of the more important should perhaps be just mentioned here. Typological theory is inextricably linked with philosophical idealism, which on pragmatic grounds (if no others!) must be excluded from modern science. The concept of distinct and static patterns cannot meaningfully be applied to real groups of organisms, which are parts of an evolutionary continuum and which are always highly variable. Their variation is not incidental or an "accident" to be ignored at any level in taxonomy; it belongs to the very nature of taxa and is part of the mechanism of their origin and continuing existence. Definition of taxa by characters in common is a special case, even though a very frequent one as a convenience in practice. When the whole four-dimensional aspect of the animal kingdom is viewed, strictly typological definition is sometimes impossible and is never a suitable theoretical basis.

THE DISCOVERY OF PHYLOGENY

Now that evolution has been discovered, we see that all the attributes used in any classification originated historically by descent with modification. It is also evident that all the taxa of pre- and non-evolutionary classification, and of course also those of evolutionary classification, likewise arose phylogenetically. Whether in given cases they are (in some sense) natural phylogenetic units is a different question to be considered later on in this chapter. With only the rarest exceptions, zoologists now agree that phylogeny is the appropriate theoretical background for taxonomy and that it is essential for understanding and explaining all the associations involved in classification. Important disagreement persists only as to the desirable or practicable relationship between phylogeny and classification, and especially whether phylogeny can and should provide criteria for classification or can and should play its taxonomic role only in interpreting classifications based on other criteria. I hold the former view, and will defend it in more detail as we go along.

Linnaeus, Buffon (1707–1788), and some other pre-evolutionary taxonomists recognized the possibility, at least, that species might not be completely static but might have changed (for example, "degenerated") appreciably since they were created. There was even speculation that the "kinds" of creation (Genesis I: 12, 21, 24, 25) might be the genera or even higher taxa of today and that species *within* each kind might have evolved since the creation. (This view is still seriously maintained by some fundamentalists who seek thus to reconcile a literal interpretation of Genesis with admission of some of the evidence for evolution, for example, Marsh, 1944.)

Lamarck (1744–1829) was, however, the first to maintain clearly and consistently that *all* taxa have arisen by evolution and are a phylogenetic continuum.[8] His idea of the continuum was far more literal and extensive than that of any modern phylogeneticist. He held in theory that there are *no* gaps in nature, even between different phylogenetic lineages, but only one continuous progression through which *all* organisms have tended to pass throughout the history of life and are still doing so today. Taxa, then, are merely arbitrary (not therefore necessarily unnatural) subdivisions of the continuum, and it is a corollary of the completeness of that continuum that no taxa can become extinct. Of course Lamarck recognized that there were in fact profound gaps between taxa actually known and represented in museums, for instance between mollusks and fishes, and like everyone else he used those gaps for the practical delimitation of taxa. Nevertheless, he was confident that the gaps would be completely filled by discovery of living organisms not then as yet found and collected.

The main implications for classification that we now see in Lamarck's theory are: first, that all taxa must be arbitrarily delimited (and cannot very well have archetypes) except as they are *temporarily* separated by gaps in knowledge, not in nature; and second, that taxa at any one level should have a natural sequence from lower to higher, simple to complex. Lamarck himself did not really develop the first implication, and the second made no difference in practice because the same practical consequence for classification was inherent in the

[8] That is the most original and the one still acceptable principle of Lamarckism. Lamarck would be astonished to know that "Lamarckism" has come to mean the inheritance of acquired characters. In the first place, he did not believe that *all* acquired characters are heritable, and in the second place in saying that *some* are, he was only repeating what everyone "knew" and had "known" since antiquity.

ancient and nonevolutionary concept of a *scala naturae* or chain of
being. In fact Lamarck's classifications did not differ in any important
way from those of his nonevolutionary contemporaries. Like most of
them, he was concerned with the problem of ranking characters, and
in this respect his views were essentially like those of his antievolu-
tionary archenemy Cuvier. "Relationships are greater according to
whether the part involved is more essential." (Lamarck, 1809, the basic
reference for Lamarckian theory.) In animals the most important relationships are derived from the parts most essential in preserving their
lives, and in plants from those most essential to reproduction (hence
not markedly different from Linnaeus's sexual system). As a matter
of historical fact, Lamarck's evolutionism did not promote and may
even have retarded the development of an evolutionary taxonomy. He
himself made no significant progress in that direction, and his views
on this subject, as far as they differed from those of, say, Cuvier, were
almost universally rejected. They were, indeed, obscurely tangled
with much that seemed absurd then and seems even more so now,
such as his belief that all chemical compounds are of organic origin.

Other claims have been made for pre-Darwinian founders of evolutionary taxonomy, but in my opinion none needs to be taken more seriously than Lamarck's, and I have been forced to disallow his claim
(which he did not make for himself). Evolutionary taxonomy stems
explicitly and almost exclusively from Darwin. The subject is recurrent
throughout *The Origin of Species*, but especially in the chapter on
"Mutual Affinities of Organic Beings." (Chapter XIII in the first edition, Darwin, 1859, and Chapter XIV in the sixth edition, 1872; in order
to follow Darwin's views in detail the variorum edition, Morse Peckham, ed., 1959, is indispensable.)

In *The Origin of Species* Darwin was chiefly concerned with estab-
lishing the truth of evolution and the effectiveness of natural selection
as the mechanism of evolution. He suggested many more detailed
ramifications that might ensue, but in this summary work he could not
be concerned with following them up. That is particularly true of
classification. His own major classificatory work (that on barnacles)
was completed not before he had formed, but before he had published
his views on evolution, and it is not overtly evolutionary. He never
returned to detailed classification (which in fact he came to dislike)

and thus did not himself apply the taxonomic principles stated or implicit in *The Origin*. What he actually did was to take as given the classifications then current and to show, first, that they were consistent with the theory that their taxa originated by evolution, and second, that evolutionary phylogeny could explain the order that had *already* been found among organisms. He submitted, with complete justification, that these demonstrations constituted the strongest kind of evidence for the truth of evolution. That, and not the remodeling of classifications along evolutionary lines, was his main purpose.

The process of evolution tends to produce groups recognizable in terms of similarities and differences. It therefore should occasion no surprise that, in considerable part, phylogenetically related groups had indeed been recognized before the real nature of their relationship was known. If that were all of it, Darwin's contribution to comprehension of taxonomy would still be tremendous, but it would have had comparatively little influence on the practice of classification. In fact Darwin did, almost incidentally, do more. He suggested new criteria or, as Cain (1959b) puts it, a priori rules that finally broke the vicious circle of Cuvierian apriorism. He also suggested and illustrated the fact that these criteria in some instances do justify the arrangements of organisms that had already been made but in other instances do not. They therefore implied changes not only in the theory but also in the procedures and in the results of classification. Although Darwin did not himself pursue the matter further, his successors have done so, and largely along lines suggested by him.

Reduced to barest minimum, the essential principles of strictly Darwinian taxonomy are as follows. Taxonomic groups are the results of descent with modification, or of phylogeny in the term now more usual. Each valid taxon has a common ancestry. The most fundamental *but not the only* [9] criterion for the ranking of taxa is propinquity of descent. Characters used in definition are to be interpreted as evidence of phylogenetic affinities and to be ranked and evaluated in accordance with their probable bearing on propinquity of descent.

The kinds of evidence explicitly discussed by Darwin were: (1) characters in common interpreted as heritage from a common an-

[9] Italicized because the incorrect statement that this is the only criterion in Darwinian taxonomy is frequently made by its critics.

cestry, (2) chains of contemporaneous taxa interpreted as having diversified from a common ancestry with survival of intermediates, (3) temporal succession of changing forms interpreted as an actual record of phylogeny, and (4) clearly inferable community of genetic origin. All of them are still used, and others have been added (see Chapter 3). As regards (1), Darwin was well aware that characters in common do not *necessarily* indicate common ancestry and that this was a serious flaw in most nonevolutionary classifications. I am thus quite unable to follow Cain (1959b) in considering this the main flaw in *Darwin's* taxonomy. It also seems to me beside the point that Darwin's taxonomy does not give apodictic certainty, Cain's major criticism in his generally favorable summing up on Darwin. It seems quite impossible that *any* system should be apodictically certain. Any certainty that may be claimed for such an empirical approach as, for example, that of Cain and Harrison (1958) seems to me to be spurious. That is only a good way of arranging evidence, the selection, the interpretation, and the formalization of which depend both on judgments of probability and on personal taste.

I have elsewhere (Simpson, 1959c) discussed Darwin's taxonomy at somewhat greater length, and the rest of this book is largely devoted to results of following up his premises. Further consideration of them is therefore unnecessary at this point.

WHAT IS NATURAL CLASSIFICATION?

Here we pause, as something of an interlude, to consider the constantly recurring question that heads this section. For two or three centuries taxonomists have been insisting that classification should be natural. Most of them have maintained that their own systems were natural, at least in principle. Others, while adopting frankly artificial approaches as Linnaeus did in some of his work, have nevertheless emphasized the theoretical desirability of natural classification. With all that, ideas as to what a natural classification actually is have been numerous and contradictory, and the expression has generally been left undefined or defined in vague and unsatisfactory terms.

Any classification that can be applied to real organisms must of course have some reference to characteristics that the organisms do actually have in nature. Using the term "natural" merely on that account would be entirely pointless. Natural classification must imply that the classification itself and as such is in some sense natural, and not only that it involves natural characteristics or criteria. The taxa of natural classification must have some relationship (not necessarily a point for point identity) with *groups* of whole organisms really existing in nature.

Idealistic-typological and special-creationist principles, which, as we have seen, often coincided and were not necessarily alternatives, postulated the existence of a unique system of archetypes, of fixed "ideas," "types," or "patterns of creation," in nature. To produce a completely natural classification it was, then, necessary only to recognize those archetypes and to group organisms in accordance with them. But that word "only" overlies a difficulty that is really insuperable: no agreement is possible as to *how* the archetypes are to be recognized, and in practice the postulated unique natural classification cannot be reached in this way. In fact it could hardly be achieved because it simply does not exist. A strict archetypic approach, involving no ulterior criteria on quite different principles, is impossible even in theory. Frequently, indeed usually, different archetypes result from different groupings which are made *before* the archetype is recognized and on some other basis, as in this simple example:

$$
\begin{array}{lccc}
\text{Species:} & \text{I} & \text{II} & \text{III} \\
\\
\text{Attributes:} & \left\{ \begin{array}{ccc} a & a & e \\ b & b & b \\ c & d & d \end{array} \right. \\
\\
\text{Archetypes:} & ab & bd
\end{array}
$$

The actual situation is rarely so simple, but this ambiguity constantly recurs in more complex forms. It is, for instance, easy enough to abstract a carnivore archetype and a marsupial archetype *after* grouping organisms into carnivores and into marsupials. To which should the Tasmanian wolf, *Thylacinus*, be referred? If only its own attributes are analyzed typologically, it could equally well belong to either or both, but one unites it with placental wolves, *Canis*, which indeed it resembles closely in most obvious attributes, and the other with kangaroos, *Macropus* etc., to which its obvious resemblances are much more distant. If, as is the case, even typological classifiers selected the marsupial archetype as more important or prior in classifying *Thylacinus*, they did so by induction after making a taxon Marsupialia (including *Thylacinus*) on other grounds and not by deduction from the archetypic principle.

The Linnaean, Cuvierian, and other systems for assigning priorities to various kinds of attributes were attempts to solve this inherently insoluble problem. It has already been shown that a circular fallacy was involved and that no revelation of a natural, unique, sequence of archetypes did or could in fact emerge from those procedures. The many discussions as to the extent of natural existence of such archetypic groups now seem rather futile, since it has become clear enough that none of them really exist as such. Linnaeus, for example, vacillated to some degree but as a rule held that his genera and probably also his species, when correctly diagnosed, corresponded with real units of creation and hence were fully natural but that higher taxa were less natural or frankly unnatural. Correct diagnosis of a species was according to its *nature*, which in turn was determined by its *essence*, which implies definition by *genus* and *differentia* in the scholastic sense, but (as already emphasized) no proper criteria for recognition of the essence were or are possible. (Again see Cain, especially 1958a.)

Another approach throughout the whole history of classification has been the idea that classification by single characters at each level or simple dichotomies, as in keys, is unnatural but that classification by simultaneous use of numerous attributes and, when indicated, multiple subsets is natural. Although often repeated by learned taxonomists, that flat statement, taken literally, verges on nonsense. It implies that if you add enough zeros (associations with *no* natural basis) you will eventually get a positive quantity (a natural association). Nevertheless the statement is not meaningless; it only omits hidden postulates and premises that are necessary to give it meaning.

One has only to examine a number of widely accepted classifications to see that the point is not so much how many attributes are taken into consideration as what ones are selected and how they are interpreted.

"Animals with hair," involving a single attribute, characterizes a group that all taxonomists have long considered natural, whatever their theoretical concept of naturalness. "Animals with wings" does not, although the one attribute here also involved seems logically on a par (equivalent in level) with the attribute of having hair. The real reason for bringing in multiple attributes is not that this is somehow more natural in itself but simply that it increases the probability of forming groups that taxonomists in general will feel and agree are more natural. It also incidentally but importantly increases the content, meaningfulness, and inductive usefulness of classification, the point made by Gilmour (1951) as quoted in Chapter 1, although he seems to me illogical when he equates these advantages with naturalness.

The feelings and agreements of taxonomists have impelling pragmatic significance but no direct theoretical value. In fact much of the theoretical discussion in the history of taxonomy has, beneath its impersonal language and objective façade, been an attempt to find some theoretical basis for these personal and subjective results. An essential part of that basis, although not all of it, is the simple fact that readily recognizable and definable groups of associated organisms do really occur in nature. In spite of doubtful cases and myriads of complications, it is quite obvious to a modern scientist, as it was to a prehistoric Guaraní Indian, that natural species do exist. The theoretical problem, then, is why they exist, and the solution of that problem must provide the criterion for whether a taxon does in fact correspond with a natural group. The general solution is now known, even though it brings with it numerous subsidiary problems, some already solved and some not. Species exist because they evolved. That, in briefest form, is the natural reason for the existence of species and is therefore also the truly natural basis for classification. The natural existence of higher taxa is less intuitively obvious, and that is the reason for all the soul-searchings of taxonomists as to whether those taxa are also "real" or "natural." But evolution does also produce higher taxa and this is the line to follow in attempting to recognize natural taxa at all levels.

In short, if such a thing as natural classification can meaningfully be achieved, it must be by evolutionary classification.

Different approaches to classification are related to different ways of formalizing relationships, of looking at them or visualizing them. In this section I shall quite briefly consider some of these ways and the most general methods of transforming them into or representing them by a hierarchy.

Relationships in general and especially associations of similarity can sometimes be broken down into conceptual units. One can visualize the presence of any one of these unit relationships as a line connecting two points that represent either individual organisms or lower taxa. When multiple points are considered, the result is a sort of network connecting some points and not others (Figure 1A). The set of connected points may then be considered a taxon, T_1 if the points are individuals and T_{j+1} if the points are T_js. Use of multiple unit relationships permits connecting all the points, some by single lines and some by two, three, four, or more (Figure 1B). Sets of points connected by many lines might then be considered lower taxa and those connected by fewer lines higher taxa. An attempt actually to do this with Figure 1B will show that it is possible if certain conventions are invented, but that it leads to some anomalous and equivocal results. For example, c is most closely related to both a and b, but as far as these attributes are concerned there is no relationship at all between a and b. The approach is typological and the fact that such anomalies, and some far more complex, do indeed arise in typological classifications of real organisms is one of the defects of that unacceptable system. Nevertheless network formalization was early proposed in the history of taxonomy and still serves for the systematization (not directly the interpretation) of some data such as the results of reciprocal crosses among genetical populations.

Another way of representing similar data has a far more direct and workable relationship with the hierarchy and for that and other reasons has been much more widely used. It is also more flexible in the kinds of relationships that it can represent. This is the group-in-group formalization, readily visualized as a series of circles or boxes of different sizes and inclusiveness (Figure 1C). Circles, for instance, of the same size can be directly equated with taxa of the same rank, and inclusion in larger circles with inclusion in higher taxa. In typological terms, the circles represent archetypes of increasing generality or abstraction. (As already emphasized, arbitrary assignments of priority to different attributes can give quite different arrangements of the same data.) By direct transfer into early evolutionary concepts, the

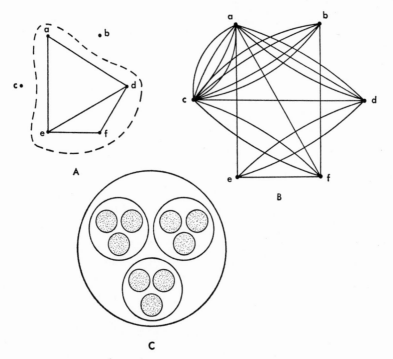

FIGURE 1. DIAGRAMS OF RELATIONSHIPS

The lettered dots represent either individuals or low taxa.

A. Representation of one unit relationship, which unites *a, d, e,* and *f* into a topological taxon.

B. Representation of multiple unit relationships; the reader is invited to form taxa on this basis.

C. Diagram of group-in-group relationships. Dots represent individuals, small circles T_1 taxa, larger circles T_2 taxa, and the largest circle a T_3 taxon.

circles delimit characters in common derived from common ancestries: the largest circle bounds characters from a remote ancestry shared by all individuals and the smaller circles characters from a less remote ancestry shared only by the individuals within each circle. The numbers of shared characters is (approximately) *inversely* proportional to the sizes of the circles because they are additive: the individuals in a small circle share not only the characters represented by that circle but also those represented by the larger circles that include it. With a more complete break from typological thinking, the circles retain the relationship to common ancestry, the remoteness of which is roughly proportional to the sizes of the circles, but they no longer have any rigid or necessary relationship to characters in common. There remains only the statistical relationship, with many exceptions in particular cases, that phylogenetically closer groups tend on an average to have more characters in common, which was of course the truth veiled behind the false typological approach.

The third formalization to be mentioned is both simpler and older than the first two: it is merely a line originally taken to represent the *scala naturae* (Figure 2). In its purest form as advocated, for example, by Bonnet (1745) there was supposed to be just one line, unbranched and unbroken, with individuals densely spaced all along it and running, in Bonnet's system, from man at the top to "subtler matter" (virtually complete formlessness) at the bottom. The system is radically nontypological and its theoretical relationship to the hierarchy is more arbitrary but just as clear and simple as in the typological groups within groups. For transformation into a hierarchy the continuous line is arbitrarily and completely divided into small segments, which are the T_1s, and these are included in successively larger, likewise arbitrary segments, T_2s, T_3s, and so on (also shown in Figure 2). No truly evolutionary concept was involved in the *scala naturae,* and yet like typology it was based on selection and misinterpretation of real phenomena that are in fact results of evolution. The formalization can therefore be transferred directly to parts of the true evolutionary pattern: it equally well represents a single, ancestral-descendant lineage of continuous populations changing gradually through time. Translation into hierarchic terms can also be made by the same arbitrary procedure of cutting into segments of varying sizes.

The point will be discussed later, but even here it is advisable just to mention that such arbitrary subdivision does not necessarily produce taxa that are either "unreal" or "unnatural," as has sometimes been stated. A simple but, at this point, sufficiently explanatory analogy is provided by a piece of string that shades continuously from, say, blue at one end to green at the other. Cutting the string into two is an arbitrary act, but the resulting pieces are perfectly real sections of string that existed as natural parts of the whole before they were

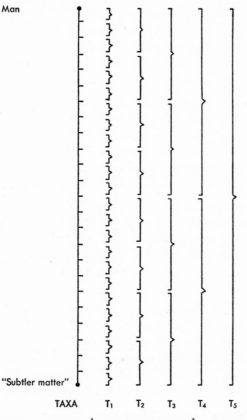

FIGURE 2. DIAGRAM OF BONNET'S CONTINUOUS *scala naturae* AND OF ITS ARBITRARY SEGMENTATION INTO TAXA OF SUCCESSIVELY HIGHER RANKS

severed. Moreover the two segments really are significantly different in average color and are naturally definable in those terms.

There are several other ways to visualize various relationships,[10] but the last to be considered here is the *tree*. It is the now widely

[10] For example, by overlapping areas to represent a key, or downwardly branching lines to represent a pedigree.

familiar pattern of multiple, upwardly branching lines. This sort of visualization seems to be implicit in the work of some pre-evolutionary authors such as Buffon, but most general use has been evolutionary because trees do seem more apt than other representations for truly evolutionary patterns. The earliest published tree diagram was probably that of Lamarck (1809), and a diagrammatic tree is the only figure in *The Origin of Species* (Darwin, 1859). In such usages the tree is meant to represent, necessarily in a diagrammatic and incomplete way, the actual pattern of phylogeny. The lines are evolutionary lineages and the vertical dimension represents time. A horizontal cross section represents the groups as they were, or are, at any one time. This cross section presents the picture that may be more or less adequately shown in a group-in-group diagram, and the tree provides the evolutionary basis and interpretation of that picture. Its groups are definable in terms of common origins shown at deeper levels in the tree, and their hierarchic ranks are determined (with some necessary approximations and restrictions discussed elsewhere) by the relative remotenesses of the various common stems (Figure 3A, C).

The tree may, however, also be used to represent relationships, even strictly typological resemblances, that are not intended to be understood phylogenetically (see, for example, Michener, 1957). The relationships are determined in a group-in-group way, and then their underlying union into the stems from which they branch are arranged so as to reflect the hierarchic ranks given to the groups so united (Figure 3B). The commonest form of such a diagram has the ends of the various branches representing taxa, while the lines leading to them do not represent known taxa but only formalize inferred relationships. Mayr, Linsley, and Usinger (1953), who have discussed this subject briefly and well, coined the term *dendrogram* for such trees.[11] Opinions or claims regarding dendrograms vary between two extremes: that they have nothing whatever to do with phylogeny or that they do faithfully represent phylogeny. Individual examples evi-

---

[11] They distinguish dendrograms from phylogenetic trees or diagrams but the distinction is not absolutely clear because their dendrograms do seem to have strong phylogenetic implications, at least. Also the word "dendrogram" would seem etymologically appropriate for all tree diagrams, and the distinction between those that are intended to be phylogenetic and those that are not could better be made in some other terms. However, I see no real need to complicate the subject by coining such terms here.

dently closely approach both extremes, but neither can be quite literally correct in any case. A dendrogram necessarily formalizes results of phylogeny and can hardly be fully independent of their cause. On the other hand, a dendrogram by definition represents at most only inferred and not observed phylogeny and therefore is unlikely to be an exact representation of the phylogeny.

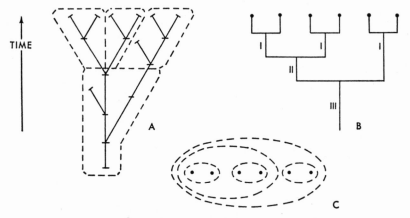

FIGURE 3. DIAGRAMS IN THE FORM OF TREES

A. A hypothetical phylogeny. All the solid lines are entirely composed of species, each of which has a length (in time) along the line as suggested by crossbars separating them. The broken lines suggest one of several ways in which the species might be assigned to genera.

B. A dendrogram with no time dimension. The terminal dots represent contemporaneous species, also without representation of their time dimension. The lines below are not composed of species but formalize relationships among the species represented by the dots. The numbered stems could be used to group the species into phylogenetically valid higher taxa. Note that although this is not a phylogeny, it is topologically identical with the phylogeny shown in A except for the omission of extinct, nonancestral species, the existence of which cannot be inferred from recent data.

C. A group-in-group arrangement of the terminal species of A or the species of B.

THE NEW SYSTEMATICS AND AFTER

General interest in zoological classification, which had been intense from about the seventeenth into the nineteenth centuries, tended to wane markedly around the end of the nineteenth and early in the twentieth centuries. In some quarters it came to be equated with mere identification and labeling of specimens, and after the first flurry of adjustment to evolutionary concepts the broader study of taxonomic principles seemed to be producing little that was new and important. Indeed some taxonomists themselves apparently felt that the principles were adequately established and all that remained was a tremendous but dull job of cataloguing. There was also a tendency for the bright young men to go into newer and more obviously exciting fields of biology, notably genetics and biochemistry.

Sometime around the 1930's there was a reversal of this trend. Broader concepts of systematics, fortunately never wholly abandoned by taxonomists, began to gain wider currency. It came to be more generally realized that systematics involves and affects the newer as much as the older branches of biology and that it is itself an exciting and rewarding source of insight and means of synthesis. Some systematists also felt that taxonomy had by no means achieved an adequate and stable body of principle but still required radical theoretical overhauling. Other systematists disagreed, and the resulting polemics (not yet stilled) were themselves a stimulating influence.

One of the landmarks of that movement, both a result of preceding change and a cause of later progress, was a symposium on *The New Systematics* (Huxley, 1940), which also gave a name to the movement.[12] As a matter of fact, from that book alone it is hardly possible to determine exactly what was new about its systematics or to draw up a formal statement of its principles, but it did make clear that there was a ferment working in the field and did point a general direction that taxonomy was to follow. Now after two decades the designation "new" has become relative, and systematics in general and taxonomy in particular have moved on considerably from the position reached in 1940. Indeed, "new" must always be relative and qualified, for science is always changing and yet each apparent novelty always has deep roots in the past.

Now it is possible to draw up a partial list of the main concepts and principles of the modern form of the new systematics, or perhaps one might say "the *new* new systematics." That will be done briefly and

[12] Although the expression "new systematics" had previously been used by Hubbs and probably others.

flatly here. Contrasts with the older systematics (which is nevertheless the source of the new) are largely evident from what has been said earlier in this chapter. The development of these principles and consideration of problems in their application will go on through the rest of this book.

Populations, not individuals, are the units of systematics and are the things classified. All populations vary, and the variation is an essential part of their nature and their definition. They do not have single fixed patterns or types. Types in a completely different sense are retained only for the strictly legalistic requirements of nomenclature.

Populations are dynamic systems that evolve, both by progressive change within them and by diversification and separation into multiple distinct new systems. The fundamental unit of evolution is the species. That category cannot be naturally defined in terms of static pattern or morphology but only in terms of dynamic, evolutionary, genetical concepts and relationships among and between populations. Observations of individual morphology and other somatic attributes are essential, but their role is not definitive in itself but is to provide evidence that the evolutionary definition is met by a particular population.

Taxonomic studies are always statistical in nature. The true object of enquiry, the population in nature, can rarely be observed directly and entire. Procedure must be by inference from the statistical sample, the specimens in hand, to the population. In particular instances, specific statistical methods and parameters are used whenever they are available, appropriate, and demonstrably useful. Even when they are not used, the viewpoint is statistical in principle.

Taxa, at all levels, are not in principle defined by membership but by relationship. That is, they cannot be defined solely by specifying the individuals that belong to them or listing the characteristics of those individuals but only by implicit or explicit specification of relationships among those individuals. Characters in common and overall similarity cannot play primary roles but are, as above, viewed as evidence of the relationships, which are primary. The pertinent relationships are evolutionary or, in a broad sense (requiring subsequent discussion) phylogenetic. The ranks of contemporaneous subsets, or $T_{j-1}s$, are primarily but not exclusively dependent on the dates of their sources or common ancestry relative to those of other subsets and of the sets, or $T_j s$, to which they belong. Among other pertinent considerations, also fully evolutionary but phylogenetic only in a broader sense, are the degrees of distinctiveness and the diversity of members of the taxa in question.

Supraspecific taxa are delimited on the principle of monophyly: all members of the taxon should have a single phylogenetic origin. The appropriate and practical criterion of singleness of origin is not derivation from one individual, pair, species, or genus, but derivation from an ancestral taxon of the same rank or lower.

All taxa have a time dimension, which is pertinent to any definition or other consideration of them. For purposes of formal classification, a continuous sequence of changing populations may have to be arbitrarily divided into earlier and later taxa at one or more levels, but the taxa thus arbitrarily delimited are not *ipso facto* unnatural.

The construction of formal classifications of particular groups is an essential part and a useful outcome of taxonomic effort but is not the whole or even the focal aim. The aim of taxonomy is to understand the groupings and relationships of organisms in biological terms.

3

Taxonomic Evidence and Evolutionary Interpretation

Modern taxonomy is evolutionary and its basis involves phylogeny, which cannot be directly observed and often must be inferred from nonpaleontological data lacking the essential time dimension. The *evidence* that taxonomic principles are met and that classification has phylogenetic validity is thus distinct from the *definitions* of taxonomic entities. The data used as evidence are still largely anatomical, but are being expanded into many other fields, the most important of which are to be mentioned and exemplified in this chapter. They include practically everything that can be known about an animal.

To form evolutionary taxa, their propinquity of descent is crucial and this is to be judged largely on similarities among them. Similarities are, however, of different kinds. They may arise by homology, parallelism, convergence, mimicry, and chance. These processes must be defined, and then their results must be interpreted in evolutionary taxonomic ways. A major problem is distinguishing between homology, which reflects propinquity of descent, and convergence, which does not. Criteria for recognizing convergence are to be considered in some detail.

Not only characters in common but also sequences of varying characters within groups enter into phylogenetic and other taxonomic studies. Here the main problem is to distinguish primitive from specialized characters in a sequence and to relate characters to the propinquity of the common ancestry.

Finally the nature of parallelism will be defined, and it will be found that no really important additional principle of interpretation is demanded by this phenomenon.

DEFINITION AND EVIDENCE

The principles of modern taxonomy are evolutionary and the approach to classification here taken is correspondingly evolutionary or in a somewhat special sense phylogenetic. Many evolutionary processes can be observed in action, both in field and laboratory, and so can extremely short segments of phylogeny. Those brief segments have great value for exemplification and for developing valid principles, but they have little practical application to classification beyond the lowest taxonomic levels, at best. Even the long series provided in many instances by paleontology are phylogenetic only by inference: the actual processes of reproduction and descent are not observed. (Even in the best of such examples, moreover, it is highly unlikely that specimens in an earlier sample include all or many of the *individual* ancestors of those in a later sample.)

It is therefore true that evolutionary classification uses, for the most part, concepts and definitions for which the data are not directly observable. This is not a feature peculiar to taxonomy. It is shared in greater or less degree by most of the inductive sciences. They are not, on that account, less scientific, nor are their conclusions necessarily any less, or more, certain than if direct observation were possible. In an analogous way, although for quite different reasons, atomic physics deals with things that have never been directly observed, but no one would question the validity or utility of its interpretations in terms of particles and processes that are known only by inference.

Scientists themselves frequently seem confused as to the degrees and indeed also the *kinds* of "certainty" (actually always of probability) that are required or are possible in science. Thus Thompson (1952), an able taxonomist, would deny even the *possibility* of determining evolutionary homology and hence (as will appear) of there being any such thing as evolutionary classification. "Who can say that the fore limb of the bat was ever anything but a wing?" One answer is that if we cannot say that, with a probability so great as to equal certainty for all practical purposes, then we cannot ascertain *any* biological principles of real interest and might as well abandon science and return forthwith to animism.

Here it is necessary again to emphasize the distinction between definition and the evidence that the definition is met. We propose to *define* taxonomic categories in evolutionary and to the largest extent phylogenetic terms, but to use evidence that is almost entirely nonphylogenetic when taken as individual observations. In spite of considerable confusion about this distinction, even among some taxonomists, it is really not particularly difficult or esoteric. The well-known example of monozygotic ("identical") twins is explanatory and is something more than an analogy. We *define* such twins as two individuals developed from one zygote. No one has ever seen this occur in humans, but we recognize when the definition is met by *evidence* of similarities sufficient to sustain the inference. The individuals in question are not twins because they are similar but, quite the contrary, are similar because they are twins. Precisely so, individuals do not belong in the same taxon because they are similar, but they are similar because they belong to the same taxon. (Linnaeus was quite right when he said that the genus makes the characters, not vice versa, even though he did not know what makes the genus.) That statement is a central element in evolutionary taxonomy, and the alternative clearly distinguishes it from nonevolutionary taxonomy. Another way to put the matter is to say that categories are defined in phylogenetic terms but that taxa are defined by somatic relationships that result from phylogeny and are evidence that the categorical definition is met.

The most trenchant criticisms of evolutionary taxonomy have been based on misapprehension of its principles or on setting up straw men. Thus Thompson (1952), again, thinks that evolutionary taxonomy argues "that the species is the ancestor of the individuals, the genus the ancestor of the species, the family the ancestor of the genera." That is nonsense, as Thompson implies, but it certainly is not a principle of evolutionary taxonomy. Other critics maintain that phylogenetic taxonomy is, or should be, based entirely on ancestral-descendant relationships, but of course phylogeneticists know (in analogical terms) that brothers are as closely related genetically as fathers and sons. Still other critics maintain correctly that classification cannot adequately express phylogeny, but in spite of some carelessness in language no capable phylogeneticist ever thought that it could. But the best answer to those who criticize evolutionary classification for what it is not is to explain what it is.

A first step is to consider the kinds of evidence from which evolutionary taxonomic conclusions can usefully be drawn. Observation is of individuals, but it may and as far as practicable should include whole individuals and their whole natural environments, both biotic and physical. Just what that means will now be considered and exemplified in more detail.

KINDS OF EVIDENCE

Classification originally developed and is still most active in connection with museums or, at any rate, collections of preserved specimens. It is only now, through the broader influence of systematics, becoming also in part an experimental and field science. The data used in classification have therefore always been predominantly, although never exclusively, those readily observable in preserved specimens. For soft-bodied invertebrates and some other groups that has generally meant the whole of a dead animal, with all of its anatomy but nothing else. In some groups it usually included much less: for mammals customarily the skins and skulls, for birds usually only the strictly external parts. In fossils, with comparatively few exceptions, only the hard parts were or are available and frequently (especially among the terrestrial vertebrates) only fragments of those.

For those practical reasons, the data for classification have been primarily anatomical or morphological [1] and generally drawn from a part, only, of the anatomy. It has been widely felt that if only the anatomical parts studied are complicated enough—provide enough

[1] Strictly speaking, a sharp distinction should be drawn between anatomy, observation and description of the structures of animals, and morphology, the theoretical science of generalizing and interpreting those observations. The distinction is analogous to that between classification and taxonomy in this book. Anatomy and morphology have, however, been carelessly used, as if they were interchangeable, by many students, including me on occasion.

separately describable characters—classification based on them will lead to about the same conclusions as if based on other anatomical or on nonanatomical characters. For instance, that has been shown to be true for hard parts as against skins in a test case for low taxa among mammals (Simpson, 1943). In certain groups, also, experienced taxonomists have found that some parts, for example the genitalia in some insects, are readily and adequately definitive for their purposes and that other parts hardly have to be considered. Such limitation of data is, however, justifiable only for closely related groups such as subspecies of a species or species of a genus, and not always there. For higher categories more data are always desirable and generally necessary. For instance, the species of ducks were well enough worked out on external characters, but consideration of other characters markedly changed the supraspecific arrangement and clearly improved its evolutionary basis (Delacour and Mayr, 1945).

It is a corollary of the fact that *organisms* are to be classified, not their characters, that the *whole* organism should be considered in all its parts and all its aspects. There are, of course, practical limits as to what can be observed. Not everything is yet known about the most studied of species, *Homo sapiens*. Much information that would be extremely useful is and always will be unavailable for extinct species. Nevertheless it is an axiom of modern taxonomy that variety of data should be pushed as far as possible toward the limits of practicability. The object of classification should be what Hennig (1950) [2] calls the *holomorph,* all the characteristics of the individual throughout its life.

A first step is to extend anatomical data beyond those usual from

[2] I shall here acknowledge a debt to Hennig (1950), which is certainly one of the most valuable books on taxonomy (as here defined) that has yet appeared. My approach is different and I had developed most of my basic views (which I do not mean to claim as original with me) before I belatedly read that book. The agreement is of course only partial, but it is substantial. It is unfortunate that Hennig's work is not more readily available to English-speaking students, and unfortunate, too, that it has no index or glossary of its complex terminology. The most general criticism—and it is one that cuts both ways—is that Hennig seems to be almost totally unaware of the vast body of English and American studies extremely pertinent to his theme. That also obviously underlies the opinion of Kiriakoff (1959), a disciple of Hennig, that "modern phylogenetic systematics seem to be quite unknown in the United States." Such a statement can only be based on grossly deficient knowledge. There are some neotypologists in the United States, as in Europe, but they are far from being either the only American taxonomists or a consensus of them.

traditional forms of museum specimens, for example in mammals to the baculum (for example, White, 1953), to the ear ossicles (for example, Krutzsch, 1954), and of course to the soft parts (for example, Mossman, 1953). A second step is to push anatomical observation to deeper levels, to cytology and especially karyology (nuclear and particularly chromosome anatomy), the importance of which has been stressed by Cain (1958b) and exemplified by Christensen and Nielsen (1955), Matthey (1952), and Patterson and Stone (1952), among many others.

Another aspect of the holomorph is that it has a time dimension. The importance of embryology for what we now call systematics was early recognized, and the pre-evolutionary work of von Baer (for example, 1828) is still basic. Haeckel's evolutionary interpretation that ontogeny repeats phylogeny (for example, Haeckel, 1875) seemed for a time to provide an infallible guide to reconstruction of phylogeny. Unfortunately, that is not true and it eventuated in disillusionment (Garstang, 1922). Nevertheless the developing organism is a part of the holomorph, and embryology is a very crucial part of the evolutionary mechanism. (See especially Waddington, 1957, not taxonomic but demonstrating the evolutionary significance.) Now in a more judicious way embryology decidedly does continue to provide essential data for evolutionary taxonomy. (Well exemplified in Orton, 1953, 1955, and her citations.)

Physiological information is not directly available in preserved specimens and so has usually been omitted from data for classification. Nevertheless physiology was early used on a somewhat broader basis by some systematists, for example by Cuvier and others in trying to set up a priori criteria for importance of characters, as discussed in the preceding chapter. Darwin (1859) was highly dubious about its use in judging propinquity of descent, and with good reason, because the a priori systems had tended to pick out physiological similarities that are (in modern terms) particularly subject to convergence. There is still some lingering feeling that physiological resemblance is to be considered analogical in general, hence (as discussed below) unreliable or misleading in classification, and that only anatomical resemblances can be designated as homological. In fact the difficulty, as often in morphology also, was in dealing with few and simple or generalized characteristics, frequently analogous, rather than with more complex and particular physiological traits, which may be as truly and recognizably homologous as anatomical features. It is indeed obvious that function, physiological, and structure, anatomical, must be and are closely correlated, which has led Blackwelder (1959) to conclude that in these frequent cases only one of these two kinds of data needs to be recorded. It would seem, however, that both are essential and so inseparable that neither can be understood without the other.

There are, moreover, numerous physiological data of great importance for taxonomy that can only be approached in their own terms and not through structure. Here, if "physiology" is used in a broad sense, belong comparative biochemical studies that are increasingly promising for taxonomic applications. In this general field comparative antigen reaction testing of sera, or systematic serology, has the largest body of acquired data (for example, Boyden, 1953, 1958), although, as with many of these approaches, some problems of interpretation are still troublesome. Serum protein and hemaglobin electrophoresis is a still newer method of gathering related comparative data (Johnson and Wicks, 1959, and their citations). Other recent biochemical studies of special interest for taxonomy include, for example, some on enzymes (Kaplan *et al.*, 1960) and on bile salts (Haslewood, 1959 a, b).[3]

To this point I have listed such data as can be gathered from individual organisms in the laboratory. Another approach characteristic of modern taxonomy, although in some measure it has always been involved in classification, is provided by relational data: observations on organisms in their relationships to each other and to their environments. At a minimum, these additional data have always included known geographic and geologic distributions. At first extremely vague, such observations have become more and more precise. Their importance is emphasized by the population approach, because a

[3] The studies cited are on metazoans, in which the biochemical data can join with a large body of other data and facilitate a better synthetic evolutionary taxonomy. In some protists classification is at present solely on biochemical data or with such simple anatomy as to be poorly characteristic. For them evolutionary classification is not yet practicable, although not impossible with future knowledge. They do not concern us in this book.

population is a group of organisms that live or have lived together in space and in time. That relationship can be adequately substantiated only by field data. In paleontology, a pioneer example of this approach, which now has many followers, was given by Brinkmann (1929), who followed ammonite populations centimeter by centimeter through a vertical rock sequence.

Data on spatial distribution not only place animals with precision in local populations and microhabitats but also more broadly in a zoogeographic framework. Pertinence to the classification of species and subspecies is richly illustrated in Mayr (1942), and relationships to higher categories in Darlington (1957). Such studies have wide ramifications, for example in the recognition of close convergence (see below), which is most likely to occur between groups in two areas distant or isolated from each other. For example, South American old native rodents ("hystricomorphs" or caviomorphs) closely resemble some African forms, and some extinct South American marsupials (borhyaenids) closely resemble the Tasmanian "wolf" (*Thylacinus*). These similarities were taken as indications of special affinities until conclusive evidence of radical isolation of South America was in hand, strongly suggesting that the resemblances might be convergent. Renewed studies of the organisms themselves then produced adequate supporting evidence (for example, Wood, 1950, and Haslewood, 1959b, for the rodents; Simpson, 1941, for the marsupials).

Ecological data have also always been involved in taxonomy in a broad way and are also now increasingly important and precise. For example, it is an established principle, with some expectable exceptions, that species at any one time and place (nondimensional species, as Mayr has called them) are ecologically incompatible. Each generally has a distinctive, unshared niche, and two or more associated species are usually rather sharply distinct. Criteria for distinguishing intraspecific and interspecific variation have thus been provided both for recent (for example, Mayr, 1942) and for fossil (for example, Simpson, 1943) animals. Ecological data provide criteria for recognizing especially lower but also higher evolutionary taxa. They also bear on other taxonomic problems such as the particularly vital problem of convergence, which involves ecological similarity but phylogenetic difference between the convergent groups. (For general background on recent animals, see Allee *et al.*, 1949; on recent and fossil marine animals, Hedgpeth and Ladd, 1957; on fossil animals, Cloud, 1959.)

Of the many specialized ecological relationships that enter into modern taxonomy, only two will be mentioned by way of examples. Many animals, especially insects, have strong food preferences, which may be characteristic of taxa at various levels even though (like any characteristics whatever) they are subject to evolutionary change. (Reviewed by Dethier, 1954, with a large bibliography.) The host-parasite relationship is frequently so strongly specific that it provides good taxonomic data for both host and parasite. It has also frequently been postulated that related parasites tend to evolve with and to remain confined to related hosts. When this has happened, as it evidently has in some instances, evidence on relationships is provided by one group for the other. Nevertheless, transfers to very distantly related new hosts seem also to have been rather frequent, and evidence for host relationships from parasites can, at present, have only confirmatory value. (For example, Vanzolini and Guimarães, 1955; also an extensive review of parasite evidence on mammalian phylogeny by B. Patterson, unpublished at this writing.)

The most conclusive possible evidence that a given individual belongs to a given species is the observation that it is living as a member of the specific population, especially if it is seen to breed with other members of that population. That is one aspect of the kinds of *directly* phylogenetic data that can be observed and need not be inferred. Such data have been little used in the past, but are likely to assume increasing importance for the specific and infraspecific categories as taxonomists take more and more to the field. A related, comparatively recent approach has been the study of the directly genetical constitution and breeding patterns of populations both in the field and in laboratories. Detailed analysis has as yet been possible only within a few species and so does not now enter extensively into classification of most groups of animals. Such studies have, however, already assumed enormous taxonomic importance as examples and as providing firm principles that can then be used for genetical inferences in populations not yet genetically analyzed. (See especially Dobzhansky, 1951; and on application of some of the principles to fossil populations, see, for example, Simpson, 1953.)

Behavioral data have frequently been viewed with suspicion by taxonomists for the same reasons previously mentioned for physiological data. Indeed first level or elemental behavior, involving such things as posture and geometrically possible limb movements, is physiological and does suffer from the same taxonomic problem as the simpler, more generalized physiological traits: it is particularly subject to convergence. That problem is, however, much less serious for second level or compound behavior, involving more or less complicated patterns or programs compounding the elemental movements, as, for example, in bird courtship. Mayr (1958; several other chapters in Roe and Simpson, 1958, also have some relevance here) has demonstrated and adequately exemplified the taxonomic usefulness of behavior at this level, where it is relational rather than individual.

Finally, more explicit recognition of the sample-population relationship has involved another marked change in the collection of taxonomic data. It was, of course, recognized even in antiquity that many animal species include sharply different forms such as, for instance, males and females or larvae and imagos. It is also an old discovery that many species are polymorphic in a stricter sense, with distinct forms at one life stage and in one sex. The need for including such different forms was rather widely, although not consistently, recognized by pre-evolutionary taxonomists. Until comparatively recently it was, however, the general opinion (with the usual exceptions) that data from one specimen of each form, or one for the whole species if there were no obviously different forms, was sufficient for taxonomic purposes. Other specimens, if collected at all, were "duplicates" and were commonly traded to other museums or collectors.

Modern taxonomic practice, concerned with *populations* and not *types*, requires as far as possible series of specimens large enough for inference as to the total variation in the population from which the sample is drawn. Statistical methods provide precise criteria for judgment of the adequacy of a sample in hand for the inferences made from it, and also for the sizes of samples necessary to draw inferences with a given level of probability, or rather of confidence, in particular instances. Of course it is not always practical to obtain a sample of optimum size. That is likely to be particularly true of fossils, even though some very large samples are now available, especially for invertebrates.[4] Small samples and single specimens are not useless, however. They only give lower confidence levels and poorer estimates of population ranges, and these deficiencies can and must be taken into account. (See Simpson, Roe, and Lewontin, 1960.) Although the inadequacy can be overcome in time, inadequate sampling is also still a serious problem as regards some neontological data. For example, interfamilial and interordinal comparisons in systematic serology still rest in considerable part on single individuals of single species in the higher taxa compared, and the results must be correspondingly insecure as bases for taxonomic inferences.

SIMILARITY AND HOMOLOGY

As has already been mentioned and will be further discussed, evolutionary taxonomy does not rely solely on associations of similarity or in principle define all taxa by characters in common. The observation and interpretation of characters in common do nevertheless play a large and essential part in evolutionary taxonomy, as they must in most systems of classification. They have other roles as well, but much of their importance is that they are one, and on the whole the chief, of the several criteria for judging propinquity of descent (Darwin's apt phrase). Propinquity of descent, in turn, is an important one of the several criteria for allocating ranks to taxa and corresponding taxonomic priorities to characteristics of organisms.

Similarities are most directly indicative of propinquity of descent if they earlier occurred in a single taxon ancestral to the later taxa in some or all of which the similarities are retained, in other words, if they have been inherited from a common ancestry. As was already well known to Darwin and has become still more evident since, this is not the only way in which similarities arise in the course of evolution. Although absent in the common ancestry, they may arise in some

---

[4] I doubt whether finances, storage space, and time or, indeed, availability in the field will ever permit gathering entirely adequate samples of species of sauropod dinosaurs, for example.

or all descendant taxa as *parallel* developments channeled by characteristics, genetical or other, of the ancestry. In that case they still have a bearing on propinquity of descent, but the bearing is less direct and may be less definitive. They may and frequently do also arise as independent *convergent* adaptations to similar ways of life in taxa of quite different ancestries. In that case they have no bearing on propinquity of descent and are a major potential source of confusion or error in evolutionary taxonomy.[5] Similarities can also arise from mimicry, in which the similarity is itself the pertinent adaptation, and from chance, in which the similarities have separate and unrelated causes. These, too, have no bearing on propinquity of descent, but in them the similarities are so superficial that they are rarely confusing.

One of the really basic problems of evolutionary taxonomy is to distinguish among these various kinds of similarities and particularly between those that are and those that are not inherited from a common ancestry. For purposes of discussing them a vocabulary is needed, and the following definitions are here adopted:

Homology is resemblance due to inheritance from a common ancestry. The similar characters involved are *homologous,* and the noun for them is *homologues.*

Homoplasy is resemblance not due to inheritance from a common ancestry. The similar characters involved are *homoplastic.* There is no current noun for them. Homoplasy includes parallelism, convergence, analogy, mimicry, and chance similarity.

Parallelism is the development of similar characters separately in two or more lineages of common ancestry and on the basis of, or channeled by, characteristics of that ancestry. Characters so developed are *parallel,* but again there is no current noun for them.

Convergence is the development of similar characters separately in two or more lineages without a common ancestry pertinent to the

---

[5] It is, however, wrong to consider this a problem peculiar to Darwinian or later evolutionary taxonomy, as Cain (1959b) and others have implied or claimed. It is a still greater problem in typological taxonomy, because convergence also produces a "type" just as common ancestry does. Typology has no criteria for choosing between the alternative or conflicting types, and so the problem is insoluble for that system. It will be demonstrated that evolutionary taxonomy does have criteria for solving the problem in workable ways and with varying degrees of probability.

similarity but involving adaptation to similar ecological status. Similarities so developed are *convergent.* (Again no noun.)

Analogy is functional similarity not related to community of ancestry. The characters involved are *analogous,* and the noun for them is *analogues.* Convergent characters are analogous insofar as the similarity can be related to function, which is usually and perhaps always the case.

Mimicry is similarity adaptive as such and not related to community of descent. It occurs when one group of organisms resembles another of different descent within the same community and when the resemblance is adaptively advantageous to the mimicking organism for any of several reasons, specification of which need not concern us here. The mimicking organism is the *mimic* and that mimicked is its *model.* There are no commonly used special terms for the characters involved.

Chance similarity is resemblance in characteristics developed in separate taxa by independent causes and without causal relationship involving the similarity as such.

These terms as here defined and used all involve the inferred causes and evolutionary significance of the similarities in question. They are interpretive terms in the field of theoretical taxonomy, although as with all valid taxonomic concepts they have highly practical applications in classification. They are not purely descriptive terms, and indeed in straight description or strict typology there are neither reasons nor useful criteria for specification of similarities different in principle. All one can do at the observational level or in a completely empirical system is to specify what objective similarities exist and their intensity or detail.

Resemblances now known to be homologies were recognized from the very beginning of comparative anatomy as a science, which may be dated from Belon (1555). By the time of Owen (1804–1892), the last great adherent of Goethean idealistic morphology and Cuvierian classification, it was more or less clearly established that there are two kinds of resemblances, different as a rule in intensity and, in some only vaguely understood way, in quality. To distinguish these Owen redefined the already old term analogy and the newer term homology. He (Owen, 1848) defined an analogue as "a part or an organ in one animal which has the same function as another part or organ in a different animal," and a homologue as "the same organ in different animals under every variety of form and function." Taken literally, these definitions are most unsatisfactory. The same organ usually has the same functions in different animals, and in such cases the only distinction is that a homologue is "the same organ" but an analogue is "another organ." Owen (1866) later compounded the confusion by replacing "another" with "a" in the definition of analogue. Then by definition most [similar] organs were both analogues and homologues, and they differed only when the same organs had different functions, hence were homologous and not analogous, or when different organs had the same functions, hence were analogous and not homologous—important special cases not actually specified in the definitions.

Even if that distinction were made clear, the concepts are unsatisfactory because they depend entirely on what is the same and what is not. In what sense can two animals share "the same organ"? And organs are rarely if ever identical in structure in different taxa, which is clearly what Owen meant when he wrote of "different animals." There are all degrees and kinds of similarities and differences among organs, and "same" and "different" are not dichotomous but vague and entirely subjective divisions of a complex continuum. Neither Owen nor any other nonevolutionary systematist produced stated, workable criteria for making the distinction demanded by his definitions. Nevertheless Owen and other experienced systematists of his day did make a distinction in practice with considerable success by intuition and personal judgment rather than definite criteria. In a remarkable number of instances they distinguished homologues and analogues much as we would do today on evolutionary grounds. It will be remembered (Chapter 2) that they also had considerable success in distinguishing evolutionary taxa on nonevolutionary grounds. The explanation is the same in both cases. Evolution does produce more or less clearly distinguishable groups that we call taxa, and it does produce resemblances more or less clearly distinguishable as homologous and analogous. The pre-evolutionary taxonomists, who included brilliant men who did superb work on their own premises, recognized many (far from all!) of these results of evolution without knowing what produced them or what objectively distinguishes them.

It was Darwin (1859) who produced the explanation and a definite criterion for homologues: they are, as defined above, organs or, more inclusively, any similarities inherited from a common ancestry. This is a perfectly clear and objective [6] definition to replace the vague and wholly subjective "same." Darwin did not himself advance this formally as a definition, but it was soon so taken by evolutionary systematists and has been adopted by a majority of them ever since. (On the history and varying definitions of homology, analogy, and many related terms, see Haas and Simpson, 1946, and Simpson 1959c.)

There have nevertheless been numerous polemics over the definitions of homology and analogy in modern times, and an urgent minority of able and experienced systematists has insisted on return to definitions essentially like Owen's. A discussion by Boyden (1947) is among the clearest and most competent and may suffice to represent that school for present purposes. Boyden and others of like opinion do not deny that inheritance of similarities from a common ancestry has occurred, or even that this is the explanation of (most of) Owen's homologues. Neither does the opposing consensus deny that Owen's homologues are (as a rule) recognizable even in nonevolutionary terms. If, as indeed seems to be the case in some of the polemics, the argument were only about which concept is to bear the name "homology," it would be quite senseless. I understand that when Boyden uses the word it is *sensu* Owen, and he knows that I use it *sensu* Darwin. I do so, in fact, only because that is clearly the present consensus; it is more likely to be understood because it is majority usage. Otherwise I would use some other term, such as "homogeny." As it is, usage seems to demand some other term for homology *sensu* Owen *nec* Darwin. A good candidate is "morphological correspondence," which has been defined by Woodger (1945) by strict symbolic logic (but, I must confess, in a way that seems to me difficult or almost impossible to apply in practical taxonomy).

[6] From previous experience I know that I must explain this application of "objective." Phylogenetic descent really occurred and it really involved inheritance of organs and other similarities of the holomorph, or rather of the information and basis for their development. That occurred as a completely objective process in nature. It is beside the point that our *knowledge* of the objective process is inferential and incomplete. The atomic analogy again will serve to clarify the point: no one reasonably denies the objective existence of atoms, although our knowledge of atomic structure is inferential and incomplete.

In fact in the midst of the polemics there is an issue that is quite apart from whether inheritance from a common ancestry should be called "homology" or something else. That issue concerns the position of the concept of homology, as here defined, in modern taxonomy and especially whether it is a desirable and practical part of the theoretical basis for classification. Crucial to that issue is whether homology, so defined, can be distinguished from the various kinds of homoplasy in most instances and with sufficient probability. With most taxonomists, I believe that it can be, and I shall attempt to demonstrate that. Too much must not be claimed: homology is not *always* distinguishable; its recognition is extremely difficult in some cases; and we are still undoubtedly making mistakes in this respect. It is even in some classifications (for example, those of most bacteria) a virtually inapplicable concept in current practice. Nevertheless in the classification of metazoans, at least, it is always theoretically desirable and usually practically applicable. Perhaps the fact that classifications by Neo-Owenians tend more and more toward those of Neo-Darwinians is evidence to that effect.

SOME ASPECTS OF BACKGROUND AND METHODOLOGY

Before discussing criteria of homology, some general and a few special aspects of the approach to such problems should be briefly reviewed. In the first place, evolutionary taxonomy must assume as a background the whole body of modern evolutionary theory, much of which is not directly taxonomic in nature but all of which has some bearing on taxonomy. Obviously, most of this must be taken for granted here. A few of the books that summarized a phase in the development of modern evolutionary theory and that seem, taken together, to be a basis for a new phase are as follows: Dobzhansky (1937, 2nd ed. 1941, 3rd ed. 1951); Heberer (1943, 2nd ed. 1954–59); Huxley (1942); Lerner (1954); Mayr (1942); Rensch (1947, 2nd ed. 1954); Schmalhausen (1949); Simpson (1944, successor 1953); Stebbins (1950). Most later developments can be followed in the pages of the international journal *Evolution* and its reviews and citations.

Second, attention must be explicitly drawn to a well-known but not always sharp dichotomy in the kinds of data available for evolutionary taxonomy. Data on recent animals have no time dimension, or one too short to be of much real use. Since evolutionary taxonomy has a time dimension, these data are deficient in an important way. They are not themselves historical, yet must serve for drawing historical inferences. That is a basic problem as regards homology, and also in many other aspects of taxonomy. The method of approach must be comparative, and conclusions must be drawn as to sequences in which we have only the final terms. Some recent animals are more primitive in some respects than others. It is therefore often possible to form a sequence among recent animals that approximates one that occurred in time. It is essential to remember that the two are not the same thing and that it is always most improbable that one exactly corresponds with the other. When properly interpreted, such sequences essentially supplement the evidence of characters in common.

Paleontological studies are often based on contemporaneous fossils, and then are just as comparative and based on just as nonhistorical data as those on recent animals. However, when based on sequences in geological time or when relatable to recent faunas, paleontological studies do have a true time dimension and the data are directly historical. In spite of deficiencies in other respects (biased samples, incomplete anatomy, no physiology, etc.), fossils provide the soundest basis for evolutionary classification when data adequate in their own field are at hand. In some large groups (for example, mammals), as regards higher taxa, at least, classifications have come to depend more on fossils than on recent animals. Nevertheless, in the animal kingdom as a whole paleontological data are insufficient to take that leading role and probably always will be. Hope for a classification of the whole animal kingdom adequately controlled throughout by directly historical data is almost certainly futile. That is notably true of the most abundant and in some respects most difficult of all the major groups, that of the insects. The groups better documented by fossils still provide comparative data and principles that are applicable to those poorly or not so documented. Neontological data also provide many of the principles required for interpretation of the fossil record. Evolutionary classification involves both neontology and paleontology, even when fossils are lacking and even for completely extinct groups. (Many of the problems of the reconstruction of historical sequences are discussed in Simpson, 1960a.)

A comparatively simple example will clarify the relationships among these various approaches to evolutionary interpretations of comparative anatomy and thence to phylogeny. Horses and rhinoceroses are evidently related in some degree. Both, at least, are Mammalia, a taxon already well recognized before the time of Linnaeus. On examination of almost any parts, such as the legs and feet, some more detailed similarities are found. Both groups have hoofs, a resemblance already recognized as important by many pre-evolutionary authors and eventually as a homology in the old, then still unexplained sense. In both the central weight-bearing axis passes through a medial toe, the only functional toe in the horses, one of several in the rhinoceroses. That resemblance was also noted by some pre-evolutionary taxonomists (probably first by de Blainville, 1816), but only vaguely, the resemblance being designated incorrectly as the possession of an odd number of toes,[7] and only as one of several conflicting typological arrangements among which there was little to choose. There are also more detailed resemblances in, for instance, the ankle bones (tarsus) and especially the astragalus. On the other hand, horses differ from rhinoceroses quite strikingly in having only one toe and in numerous other respects, such as their thinner and more hairy skins. In those and some other characteristics rhinoceroses resemble hippopotamuses more than they do horses. One could then form several archetypes all of which were in fact adopted by one school or another of pre-evolutionary typologists. "One toe" and "multiple toe" archetypes placed the horses under one archetype and rhinoceroses and hippopotamuses under another quite different archetype. That was the most common arrangement by early taxonomists. "Thin skin" and "thick skin" produced the same collocation as far as these three groups are concerned, and that was Cuvier's arrangement. Still other archetypes gave other groupings. Linnaeus, for instance, placed the rhinoceroses with the rodents in the "Glires" and horses and hippo-

---

[7] The error is embodied in the name Perissodactyla still given to this group. There are perissodactyls with even numbers of toes and artiodactyls (etymologically, "even toes") with odd numbers of toes.

potamuses together in the "Belluae" under "Ungulata," from which rhinoceroses, although hoofed, were thus excluded. (In all these archetypes other characters in common additional to the one I have noted for each could be adduced to support the classification.)

Only with the development of evolutionary taxonomy was it firmly established that the cited horse-rhinoceros resemblances are homologies and that the other resemblances cited are not. It could then be said that de Blainville's archetype did also occur in a common ancestor and the others did not.[8] The horses and rhinoceroses then belong to some taxon, at a rather high level, from which hippopotamuses are excluded, and that arrangement soon became universal in classification, the previously more popular Cuvierian and other arrangements being forgotten except by historians. That conclusion depended on consideration of all anatomical parts of horses and rhinoceroses and their comparison with all other hoofed mammals. In the limbs, it was found that horses and rhinoceroses (also tapirs) are similar in a complex functional system and that this resemblance is accompanied by others throughout the anatomy. Hippopotamuses, on the other hand, have an equally complex but different arrangement, shared in essentials by cattle, antelopes, pigs, and many other hoofed mammals. It happens that the difference in limb function is more clearly reflected in the astragalus than in any other one part. The distinctive astragali of perissodactyls and artiodactyls were noticed by various typologists, but none of them gave this character top priority—not even Owen, who, on other and partly incorrect grounds, did endorse the arrangement later found to be evolutionary.

In this case the homologies were established without any important support from fossils, which did later prove those homologies to be correct beyond any doubt. Cuvier already knew some fossil forms that could greatly have clarified the issue if he had interpreted them as more primitive relatives of the living animals, but he did not. (His *Anoplotherium* belongs to the same superfamily as the hippo-

---

[8] In such instances it is sometimes said that the typologist, such as de Blainville, anticipated the evolutionary conclusion. That is true only as a coincidence. With so many different archetypic systems it would be extraordinary if one did not, even by pure chance, happen to correspond with the evolutionary facts even though in no way based on evolution. De Blainville's arrangement of these particular animals just happened to be the lucky one.

potamuses, and his *Palaeotherium* is related to both horses and rhinoceroses.) *Hyracotherium* (the valid technical name of "*Eohippus*") is near the common ancestry of horses and rhinoceroses and it does indeed have the characters identified as homologous in those two groups. Nevertheless, it is quite different from any recent horse or rhinoceros and is by no means intermediate between the two, and that brings up another point.

Early evolutionists, including Darwin, were still strongly affected by earlier typological habits of thought. Many of them, but not including Darwin, who had more insight, tended to think of the archetype as equal to characters in common and as corresponding with the common ancestor or as being an adequate characterization of it.[9] (They rarely used the word archetype but commonly used the concept.) It now seems that a moment's thought would reveal the fallacy. Characters in common in horses and rhinoceroses do not include numbers of toes, but their common ancestor must have had some definite number of toes, not necessarily the same number as in any recent descendant. Such considerations bring in another method of determining propinquity of descent and of characterizing a common ancestry from comparative data, a method quite different from that of characters in common. Without going into details, it may be noted that its own anatomy in comparison with that of other ungulates and mammals in general indicates that the ancestry of the recent horses must have had more than one toe and that the common ancestry of horses and rhinoceroses must have had at least as many as three and not more than five. Consideration also of the other living perissodactyls, the tapirs, makes it highly probable that the common ancestor of all perissodactyls had four toes on front feet and three on hind feet, and paleontological data prove that to be correct. The rhinoceroses (still more the tapirs) are thus in this respect structurally more primitive, or in closer propinquity to the ancestor, than the horses. It does not, however, follow that rhinoceroses are in most or in an average of respects more primitive, and that is a point requiring further careful consideration in each such case.

[9] That naïve viewpoint doubtless is the basis for such absurd imputations as the one previously quoted that evolutionary taxonomy considers the genus ancestral to its species. That is a strictly typological and not a modern evolutionary error.

In spite of the various collocations of the typologists, horses are not so convergent toward any other living group that a modern taxonomist would be likely to mistake their homologies. There is, however, an extinct South American ungulate, *Thoatherium*, between which and horses there is strong convergence, notably in the fully one-toed feet. The discoverer of *Thoatherium*, Florentino Ameghino, an evolutionary taxonomist capable in his time (1854–1911), concluded that the resemblances are homologous and that litopterns (the group to which *Thoatherium* belongs) and horses had a common ancestry closer than that, for instance, between horses and rhinoceroses. The mistake was by no means egregious, because litopterns and horses did have a common ancestry not much older, and the later convergence between these two groups is one of the closest known. Nevertheless further study, using many of the criteria next to be considered, showed that the particular resemblance in the feet is convergent, that the common ancestry of litopterns and horses had five toes, and that it was older than the common ancestry of horses and rhinoceroses.

SOME CRITERIA OF HOMOLOGY

The establishment of homology is never absolutely one hundred per cent certain, but to insist on that fact as a deficiency of evolutionary taxonomy would tend to negate the value of any and all scientific endeavor. Some homologies, for instance between the humeri of men and apes, are as nearly certain as that the sun will rise tomorrow. Others have lower degrees of probability. Some that seemed nearly certain when inadequately studied have turned out to be almost certainly not homology but homoplasy, as in the example of the litopterns and horses. The same example shows that gathering more data and applying more numerous and precise criteria can be sufficiently definitive even in such difficult cases.

As has been shown, the concept of homology grew out of the observation of characters in common. Homology does always involve characters in common, but it has also been sufficiently shown that the mere existence of characters in common or the possibility of abstracting an archetype or, its modern synonym, a morphotype is not a sufficient criterion of homology. There are also instances of established (that is, highly probable) homologues, for example, between endostyle and thyroid gland, in which the characters in common seem so few or trivial that an abstracted archetype might seem rather to oppose than to support their homology.

As far as characters in common are concerned, two criteria are fairly obvious: minuteness of resemblance and multiplicity of similarities. In both cases more analyzably distinct unit characters of the taxonomist are brought into the comparison, and it is a sound principle of all taxonomy that conclusions on affinities (which means largely on homologies) are stronger the more the characters involved. The probabilities are cumulative (although not by simple mathematical addition), and many low probabilities taken together may produce a high probability. In this connection, useful but sometimes unfortunately laborious statistical methods have been worked out for bringing multiple characters into a single comparison. Examples of such approaches have been supplied by, among others, Stroud (1953), Michener and Sokal (1957), Cain and Harrison (1958), and Jolicoeur (1959). Two special problems, not yet fully solved, arise. First, unless other criteria are also used, homoplastic as well as homologous similarity may well be involved. Using large numbers of characters probably does tend to reduce the effect of homoplasy on the result, but some effect is always possible and it may be so great as to invalidate the result. Second, it is spurious to think that using multiple characters increases probability if the characters are closely correlated either genetically or functionally. Characters that are perfectly or highly associated certainly should not be considered independent items of evidence, but only as one item. One suggested remedy (Michener and Sokal, 1957) is to use a great many apparently different characters in the expectation that any internal correlations will then have insignificant effects. That is probably true in most cases with very large numbers of characters, but it apparently has not been empirically validated and the failure to eliminate a known cause of error is logically objectionable. Another possibility is to measure correlations among all the characters and then to use strongly covariant groups as single characters. An approach toward forming such groups has been made by Olson and Miller (1958), but the procedure is extremely laborious, prohibitively so without a large electronic calculator, and the results so far achieved are, in my opinion, of questionable statistical significance.

Intricate adaptive complexes are unlikely to arise twice in exactly the same way, hence to be convergent in two occurrences, and the probability of homology is greater the more complicated the adaptation and the closer the identity. On the other hand, similar adaptation with differences in characters not requisite for the adaptation as such is a strong indication of convergent homoplasy and opposed to homology. For example, Inger (1958) has shown that in tadpoles adapted to life in torrents simple streamlining and oral suckers have evolved convergently in several phylogenetically different groups, but that characteristic details of the oral disks permit recognition of the phylogenetic groups within each of which those details are homologous. He concludes that, "If only complex adaptive structures are considered, convergence will be detected every time," which may be a little overoptimistic but not seriously so.

Characteristics of radically different kinds are as a rule unlikely to be genetically or functionally correlated, and their concordance may be conclusive in distinguishing homology from convergence. For example, in the South American rodents previously mentioned there are available zoogeographic data, anatomical data, and biochemical data (with no apparent anatomical correlation). Each of these bodies of data is equivocal or insufficient in itself, but their combination makes it extremely improbable that resemblances to Old World hystricomorphs are homologous.

It is now realized that embryos do not represent or repeat ancestral adult structure. It is nevertheless true that developmental patterns are inherited (indeed it is they and not adult patterns that are primarily governed by heredity), and that they can give excellent evidence on homology and convergence. Embryos of phylogenetically related forms are commonly more similar than adults, a fact noted by von Baer even before the phylogenetic implications were known. The opposite can nevertheless occur: embryos are sometimes less similar than the corresponding adults. One must then find other criteria to decide whether (a) the adult resemblances are homologous and the embryonic differences divergent adaptations specifically for that period of life, or whether (b) the embryonic differences indicate different phylogenetic origins and the adult resemblances are convergent. On such relationships among embryos and adults, see the extensive review by de Beer (1951). As regards particular organs, similar embryological origin is strong evidence of homology. That was also noted before the real nature of homology was known. Owen (1848) cited "a close general resemblance in the mode of development" as a criterion of homology.

It would seem that evidence of chromosomal and gene identities would be the most conclusive possible criterion of homology. In many instances that is certainly true, but it is subject to important limitations. In some groups that have been intensively investigated genetically (for example, *Drosophila*, see Patterson and Stone, 1952), phylogenetic classification on directly genetical evidence is, if not yet fully achieved, at least a clearly achievable goal. Those are, however, useful special cases and the method is not widely practicable. It is extremely improbable that we will ever have adequate genetical information on, for example, elephants or deep-sea animals. Even without direct knowledge of genetic constitution, successful hybridization is practically conclusive proof of chromosomal equivalence and secondarily of the homologous nature of structural similarities. There is a scale from complete fertility between the parents and among the offspring through well-developed but sterile F_1 offspring and various degrees of developmental failure in F_1 to complete parental sterility. Placing on that scale evidently corresponds, as an approximation, to degree of chromosomal resemblance. Such evidence is, however, almost completely confined to infraspecific and specific levels,[10] where homologies are usually recognizable anyway on other, more readily applicable criteria.

Although such a definition has been proposed on theoretical grounds, it is certainly not practical in taxonomy to define homology as resemblance caused by identity of genes. That criterion could rarely be applied where needed. In fact I consider it as also undesirable theoretically. Few variations in structures or other characteristics most useful in taxonomy, especially above the specific level, can be clearly related to individually identifiable or single genes. Most of them are significantly affected simultaneously by multiple genes and by

[10] Supposedly intergeneric hybridization, usually with sterile offspring, is possible among animals, for instance, in mammals, the artificial crosses *Bos* × *Bison*, *Equus* × *Asinus*, and *Ursus* × *Thalarctos*. In my opinion, however, this might better be taken as basis for uniting the nominal genera. I would not now give generic rank to *Bison*, *Asinus*, or *Thalarctos*.

gene interactions. It is extremely unlikely that all the genes relevant to homologous characters are identical in an ancestor and a taxonomically distinct descendant. In any case such identity probably can never be established in an operationally useful way. Moreover it is now known that virtually identical or closely similar somatic characteristics may have quite different genetic basis. What matters in taxonomy, and indeed in evolution as a whole, is the working organism as such and not primarily its genes. If a given characteristic is continuously present in an ancestor and in all the descendants of a given lineage, then it is homologous throughout even though the genetic substrate has changed. Of course this does not alter the fact that evolution of all characters does always have a genetic substrate. It only points out that genetic evolution and somatic evolution are not identical or precisely parallel and that it is somatic evolution that is more directly pertinent in taxonomy.

Ecology has special importance here because of the close connection between adaptation and homology or convergence. Although we speak somewhat loosely of useful attributes of an organism as adaptations, adaptation is not strictly a characteristic of either organism or environment but a relationship between the two. Such relationships are ecological by definition. Homologous characters arise, as a rule with some probable exceptions, from adaptation in the ancestry and are retained in descendants, hence become homologues, because they continue to be adaptive for them or, at any rate, have not become radically inadaptive. Convergence, strictly defined, involves adaptation to ecologically similar situations by two groups of distinct ancestries, at least one of which did not have the adaptation common to the convergent descendants.

Thus when the actual ancestries are not directly known, convergence is to be suspected when groups resemble each other only or most closely in some way specifically adaptive to a particular, shared ecology and are otherwise different. For example, in the classical case of *Thylacinus* and *Canis*, the resemblances, although many and detailed, are all related to a particular pattern of predatory adaptation, and in characteristics not related to that adaptation the animals are quite different. The resemblance is manifestly convergent. On the other hand, dogs (Canidae) and the panda (*Ailurus*) are markedly dissimilar in characters related to their ecological differences but are similar in many characters not so specifically adaptive. Those resemblances are (for the most part, at least) homologues. Detailed homoplasy is especially likely in groups that occur (or that originated) in different regions more or less isolated from each other, as is true of *Thylacinus* and *Canis* and numerous other pairs of ecological vicars. Such a geographic pattern may, however, also develop in groups that do have common origin and homologous adaptation, such as the Asiatic and American tapirs. Usually, as in that example, the homology is fairly obvious on other grounds. All such cases must be interpreted in the light of what is known about adaptation as a general phenomenon of evolution and not only in its taxonomic aspects (see, for example, Simpson, 1953; Underwood, 1954).

Above the specific level, at least, the most direct criterion of homology is of course the discovery of the common ancestry or of fossil lineages clearly converging backward in time toward that ancestry. When available, such evidence is conclusive or nearly so, and its bearing is fairly obvious. It must, however, be clear by now that the neotypologists are wrong when some of them maintain that homology (or phylogeny) cannot be adequately established in the absence of fossil evidence. Other criteria, among them those that have now been discussed, almost always can suffice to recognize homology with a satisfactorily high degree of probability. A striking example is provided by the extraordinary homology between mammalian ear ossicles (malleus and incus) and reptilian jawbones (articular and quadrate), which was first recognized on mainly embryological evidence in recent animals and later confirmed, with some filling in of detail, by fossils transitional between the two classes.

NOTE ON SERIAL HOMOLOGY

Owen (1866) recognized three kinds of homology: general, special, and serial. General homology was considered the correspondence between the structure of an actual organism and an abstract archetype, a concept of no present taxonomic value. Special homology corresponded more or less to homology in the modern sense, but without the explanatory criterion of common ancestry. Serial homology, a term and concept still in use, is the anatomical correspondence among repetitive or serial structures within a single organism, such as consecutive vertebrae, feathers, or leaves on a tree. It seems fairly obvious (although it has been disputed) that this is a completely different phenomenon from that of homology among different organisms. It might therefore better be called by Bronn's term, "homonomy" (see Remane, 1956, another exceptionally valuable book on taxonomy in general).

A homonomous series or system *as a whole* is usually homologous among different organisms of common origin, as hair among mammals or leaves among trees, but any one hair or any one leaf is not homologous to one on another organism. Such resemblances in the whole system have taxonomic value because they are homologous and not because of their internal homonomy, which has no special *taxonomic* significance. There are special intermediate cases. For instance, the seven cervical vertebrae almost constant in mammals can (with minor exceptions) be homologized both individually and as a series within the class, but the homology with cervical vertebrae in amphibians and reptiles, which may have quite different numbers of cervicals, applies only to the series as a whole, or the metameric field producing it. (It is further confusing that the homology of series in different animals is not what is meant by serial homology, another reason for preferring the term homonomy.) There are also some doubtful cases. For example, *Otocyon*, the big-eared fox of Africa, has four molars instead of the usual three or less in placental mammals and, without serious doubt, in its own ancestry. It has been assumed that one has been added at the end and that the first three are homologous with the usual and ancestral three. It seems at least equally likely, however, that only the molar *series* are homologous and that in one case its developmental field has produced four teeth, in others three, without strict *individual* homologies.

SEQUENCES AND SOME CRITERIA FOR PRIMITIVE AND
SPECIALIZED CHARACTERS

It has been mentioned that the evolutionary interpretation of taxonomic data depends not only on characters in common, which are all that strict empiricists usually intend to take into account, but also on other approaches and notably on the interpretation of sequences. As regards sequences of related fossils, with a real time dimension, the bearing is comparatively simple and direct and needs no detailed consideration here. The interpretation of sequences of contemporaneous animals, or other sequences without a useful time dimension, is more difficult and does require special discussion at this point. One aspect of the approach was clearly seen by Darwin (1859), who wrote:

There are crustaceans at opposite ends of the series which have hardly a character in common: yet the species at both ends, from being plainly allied to others, and so onwards can be recognized as unequivocally belonging to this, and to no other class of the Articulata.

It will also be remembered (from Chapter 2) that even some pre-evolutionary taxonomists recognized in principle that taxa may be recognized and defined by balances or chains of resemblance regardless of characters in common and without having abstractable archetypes. It was also mentioned in passing that Beckner (1959) has introduced virtually the same approach into set theory, which in its usual form is more rigidly typological. He does so by means of what he calls "polytypic concepts," a term that I find unfortunate on two grounds. First, it links this approach to typology, whereas it can and should lead away from typology and toward a more realistic and useful consideration of organisms. Second, the term "polytypic" is already in general use in taxonomy with a sharply different meaning: for a T_{j+1} taxon that includes more than one T_j. (The implication that each T_j has or is a type is of course also objectionable in that usage.) I shall not, however, coin a substitute for Beckner's term, because I propose to carry the concept further in modified form and under a different term. Beckner's definition of a "polytypic aggregation" is as follows (paraphrased to avoid use of special symbols not further needed here).

In a "polytypic aggregation":

1. Each individual has a large but unspecified number of a set of properties occurring in the aggregate as a whole.
2. Each of those properties is possessed by large numbers of those individuals.
3. No one of those properties is possessed by every individual in the aggregate.

The last condition is necessary only to distinguish such a definition of an aggregation, eventually a taxon in biological classification, from one defined typologically. It does not exclude the possibility that the same taxon could also have characters in common or that a definition of the taxon (in this case conceptually a different kind of aggregation) could also be framed in those terms. In fact one can readily form aggregations logically similar to Beckner's "polytypic aggregation" without applying the first two of his conditions strictly, as the following example shows:

$$
\begin{array}{cccccccc}
T_js: & 1 & 2 & 3 & 4 & 5 & 6 & 7 \\
& a & b & c & d & e & f & g \\
\text{Properties} & b & c & d & e & f & g & h \\
& c & d & e & f & g & h & i \\
\end{array}
$$

Aggregation or T_{j+1} —etc.

That is an aggregation strongly united and readily definable by the fact that each individual T_j possesses a majority of the properties of the next in sequence, but each individual T_j possesses only a minority of the whole set of properties and each property is possessed only by a minority of the individual T_js. This model corresponds with Darwin's view of the Crustacea, and in usually more complex form such sequences occur over and over again in biology. This model defines what I shall henceforth call a *sequence,* bearing in mind that real zoological groups in which sequences occur may and in fact commonly do also have characters in common that are not pertinent to discussion of the sequences as such and also that quite different sequences of properties commonly occur in the same organisms.

For purposes of discussion it is assumed that the "properties" or characters in common between successive members of a sequence are known to be homologous. That implies that any pair of successive members had an ancestry in common. But if, for instance, 1 and 2 and 2 and 3 had an ancestry in common, then 1 and 3 must also have had an ancestry in common, and it follows that the whole aggregation must have had an ancestry in common and may be formalized as a phylogenetically valid taxon, which was Darwin's implicit conclusion for the

Crustacea. Thus a perfectly reasonable and indeed, if the internal homologies in the sequence are correctly established, quite inescapable phylogenetic conclusion can be drawn without (or regardless of) any characters in common and without considering what the common ancestry of the whole group was actually like. Nevertheless we would like to know what the common ancestry was like. More important for classification, we may have to know something about that ancestry in order to make inferences about phylogenetic relationships of one sequence-taxon with those of another taxon that may have arisen from the same ancestry but with evolution of a different set of properties. For example, the artiodactyls form a sequence as regards certain of their properties, such as the numbers of toes, or a whole set of partly concurrent partly different sequences as different properties are taken into account, and the perissodactyls have another such complex of sequences. Our views as to relationships between artiodactyls and perissodactyls (also all other higher taxa of ungulates) will be strongly affected by inference as to whether, for example, their common ancestry had one, two, three, four, or five toes. (The known answer happens to be five, and with other evidence this implies a comparatively remote common ancestry within the whole cohort of ungulates.)

For consideration of any one ancestral character, the sequence involved as evidence must of course have a single *fundamentum,* to borrow the scholastic term; that is, it must be a sequence of alternatives regarding the same kind of character, such as numbers of toes or venation of wings. It is clear that only one alternative and not all of them could have occurred in the ancestry, and it is then a basic assumption of this approach to phylogenetic study that one of the alternatives in the sequence is either identical with or closest to the condition in the ancestry. What is wanted then, is criteria for selecting that one, which in Hennig's (1950) complex and idiosyncratic terminology is a "plesiomorph," alternatives less like the ancestry being "apomorphs."

There are a few sequences, unfortunately very few, in which a necessary direction of evolution is more or less fixed a priori. For instance, in *Drosophila* it is a mechanical necessity that certain chromosome mutations occurred before others (Dobzhansky, 1951). Much more numerous are instances in which information extraneous to the sequence establishes a probable direction. For chromosomes, again, known processes of duplication and fragmentation make it much more likely that high chromosome numbers were derived from lower, and not the reverse. Both recent sequences and paleontological data have established various trends so common within large groups of animals that they can be used, if only tentatively, for the interpretation of other nontemporal sequences. For example, among reptiles and mammals, with no known exceptions, the number of normal toes is either five or has been sequentially reduced from that number, so that it can always be assumed with sufficient probability that a smaller number has been derived from a larger. That rule of reduction applies widely to serial structures, although for some of them exceptions are frequent. Placental mammal dentitions almost always tend to be sequentially reduced from the primitive formula $\dfrac{3 \cdot 1 \cdot 4 \cdot 3}{3 \cdot 1 \cdot 4 \cdot 3}$, but known exceptions are common among cetaceans and also occur among edentates and carnivores. Reduction in number of vertebrae is also a common trend in vertebrates, but with many exceptions, notably among elongate fishes and snakes and also occasionally in other groups, even mammals. An over-all trend toward larger body size is so common, especially among vertebrates but also among some invertebrates (Newell, 1949), that some taxonomists have assumed that a larger animal is *ipso facto* excluded from the ancestry of a smaller one. There are, nevertheless, so many exceptions that the degree of probability is only moderate without collateral supporting evidence. Rensch (especially 1954) has listed and analyzed many such rules or usual sequences, and he maintains that they permit not only inferences as to ancestry but also predictions as to their descendants (Rensch, 1960).[11]

Perhaps the most common assumption about evolutionary sequences is that they tend to proceed from the simple to the complex. Such an over-all trend has certainly characterized the progression of evolution *as a whole*. An amoeba, tremendously complex as it is, is obviously simpler than an ostrich and certainly more closely resembles their common ancestor. Within particular groups of animals, however, this criterion is rarely of much real use. The *scala naturae* does not actu-

[11] Confirmation of such "prediction" can come only by discovery of an organism that has already evolved and not from an event still in the future when the "prediction" was made. There would seem to be some logical question as to whether this is in fact prediction. Should it perhaps be called "postdiction"?

ally exist, and in dealing with divergent branches the concept "simpler" may be quite irrelevant. Is an ostrich simpler than a kangaroo? Neither is in this respect determinably closer to their common ancestry, there is no really pertinent sequence among recent animals, and the question does not seem to make sense. Frequently degrees of simplicity depend on subjective definition and may be prejudged on quite extraneous grounds. In the abstract, it would seem to be simpler to have few skull bones, as in mammals, rather than many, as in the fishes ancestral to them, or to have one toe, as in *Equus*, rather than several, as in *Hyracotherium*. Again the concept of simplicity may be irrelevant or may demand special and peculiar definitions to fit the facts established on other grounds. There are, furthermore, many instances, especially among parasites, where organisms objectively simpler have been derived from more complex ancestors.

A serious problem is that almost any aggregation of varying objects can be arranged in a graded sequence as regards some of its characters, but the sequence does not necessarily have a true relationship with phylogeny. That fact underlay the origin of the now generally discarded (but still with diehard supporters) concept of orthogenesis, which is often considered paleontological in origin but really arose from misinterpretation of arbitrary, nonphylogenetic sequences among recent animals (see Jepsen, 1949). As a simple example, the numbers of toes of ungulates can be arranged in sequence 5-4-3-2-1. It is true that one end of the sequence (5) is primitive and the other (including 2 and 1 equally) specialized, but few if any one-toed ungulates have gone through an ancestral sequence 5-4-3-2-1, or two-toed ungulates through the sequence 5-4-3-2. It may also be true that neither end of such a sequence is more primitive than the other. It is, for instance, improbable that either white or black skin is primitive for man, and in some sequences no step may be demonstrably more primitive than another.

In many otherwise difficult cases, criteria similar to those for characters in common can often be developed from the distribution of characters. Among recent ungulates, for example, almost all have multiple toes and only one genus, *Equus* (*sensu lato*) has one toe. Among mammals as a whole those with five toes are not in a majority, but that number does occur in some members of most of the orders, and within each order sequences suggest reduction from five or at least four. Confirmation from fossils was not necessary to indicate that five is the ancestral number.

The terms "primitive" and "specialized" have already been introduced as essential to this discussion, and their varying usages by taxonomists have been confusing. Their usage here should, therefore, be more carefully defined before proceeding.

The concepts of primitive and specialized are relative and are meaningless unless definitely related to a particular taxon, lineage, or phylogeny.

Within a taxon, the characteristics of the common ancestry are primitive and others are more or less specialized in proportion to their departure from the ancestral condition.

Within a single lineage, characteristics occurring earlier are more or less primitive and those appearing later more or less specialized in proportion to their times of appearance.

Within a phylogeny (which is divided into taxa and consists of branching lineages) *characteristics of any one common stem are more primitive than different characteristics of its descendant branches; within single branches they are more specialized in accordance with earlier or later appearance; and between branches they are more primitive or more specialized in accordance with less or more departure from the stem condition.*

The concept "generalized," often confused with "primitive," is almost always ambiguous, and I would prefer to avoid its use in this connection. It sometimes refers to an abstraction of characters in common, which is not, or is only coincidentally, represented as such in a real organism. At other times it has the quite different application to a primitive condition that is also less narrowly or specifically adapted to a particular niche than specializations derived from it. Primitive animals are frequently relatively generalized in that sense, but the a priori expectation is not high enough to make this an acceptably probable inference without other evidence.

In contemporaneous sequences, more primitive characters will be those that have evolved more slowly. That relationship has spurred search for kinds of characters that tend to evolve slowly and that hence may be assumed to be a comparatively reliable indication of a common ancestry. No such fully reliable touchstone has been found, but some criteria for probability are available. It used to be commonly anticipated that adaptive characters were less and nonadaptive characters more reliable in this respect, because the latter were assumed to be less labile in response to varying ecologies. That criterion is now generally abandoned. In the first place, truly nonadaptive characters seem to be extremely rare and are seldom if ever certainly identifiable as such. In the second place, characters that may be nonadaptive or that are less clearly and precisely adaptive seem as a rule to be more rather than less labile. A tooth, for instance, that no longer occludes or that is in the process of reduction and loss of essential function (like human wisdom teeth) is demonstrably exceptionally variable in both structure and size.

Characteristics that are more general in nature or that represent a broad improvement in a variety of ecological situations are likely to be more primitive. That underlies Gregory's (1951) concept of "heritage" as opposed to the more specific adaptations of "habitus." In cetaceans, for example, loss of hair, streamlining, and flippers and flukes are general adaptations to fully aquatic life and are a "heritage." It is highly probable that they are primitive for the order. Such characters as whalebone, ventral grooving, tusks, and so on are "habituses" in the various lower taxa of the Cetacea and doubtless arose later. Relationship to characters in common over more or fewer taxa is evident. It is plainly a mistake to consider the "heritage" as nonadaptive and the "habitus" as adaptive. The whale "heritage" is even more essentially adaptive than any one whale habitus. It must also be remembered that broadly adaptive "heritage" characters are just those most likely to be convergent between different groups. Darwin already noted, in other terms, that the characters common to Cetacea are homologous within that group but are convergent between cetaceans and other aquatic vertebrates.

We have already noted that intricate, functionally coordinated structures are less liable to close convergence, and an aspect of that fact is that they tend to be less labile and to retain ancestral conditions longer. The ear region in mammals is such a structure, characteristically with delicate *coadaptation* among its parts. It is probable in all such cases that there is a correspondingly complex and delicate coadaptation in the genetic substrate. Almost any change then requires difficult changes throughout the complex and will be rejected or only slowly coadapted by natural selection. That evolutionary principle explains the taxonomic observation that such complexes tend to be conservative, that is, relatively little labile. The lability is never zero, however. Ear structure in mammals is a particularly "good" taxonomic character for higher categories, that is, one more likely to bear on common ancestry, but classifications based on that region *alone* are not satisfactory.

Serological reactions also seem to be relatively little labile, which is the reason for the great taxonomic interest of those data. The theoretical basis is not wholly satisfactory, because the evidently very numerous individual biochemical compounds involved have not been identified, and it is not wholly clear why their combinations should be conservative. Validation is largely empirical, from the fact that degrees of serological similarity usually do correspond closely with judgments of affinity previously made on other grounds. There is, however, danger of circular reasoning in numerous procedures so validated and not in serology alone. When the two do not agree, is this to be taken as weakening the validation or as revealing an error in the interpretation of the data used for validation?

Experience with particular groups always leads to empirically based judgment that some kinds of characters are more labile than others, and every specialist in classification acquires a "feel" for the less labile or "more reliable" characters in his group. One basis is the direct observation and quantification of degrees of variability within populations. Then as taxa are built up by coordination of all evidence, it becomes apparent that some properties do in fact characterize lower taxa, hence may be inferred to be more labile, and others, higher taxa, hence are less labile and more likely to be primitive for the whole group in question. These properties differ from group to group and must as a rule be separately determined for each, and the same properties may have quite different lability in different groups. Characters of one kind (or with the same *fundamentum*) do not have a fixed taxonomic rank; there are no kinds that are inherently of generic or familial or ordinal value, as some taxonomists used to think, or hope. For example, the number and form of incisor teeth are highly labile and usually characteristic of genera in Eocene prosimians, but are quite stable within families in recent prosimians, and (with expectable lesser variations) in the whole order of rodents. There are nevertheless a few kinds of characters, notably those of color and color pattern, that are almost always highly labile wherever encountered. We are most unlikely ever to know whether eohippus was striped, spotted, self-colored, or something else.

Perhaps the most conclusive evidence as to primitive (and hence ancestral) characters is provided when one condition in a group or one end of a sequence has a homologue in another group of more remote common ancestry. Among mammals, for example, the monotremes, marsupials, and placentals have three fairly distinct and characteristic modes of reproduction. That of the monotremes is, in several respects, clearly homologous with that of most reptiles. It is, therefore, clearly most primitive among mammals although much the least common among them. It used to be assumed that monotremes–marsupials–placentals form a sequence in reproduction, and that marsupial reproduction is therefore more primitive than placental. However, marsupial reproduction does not particularly resemble that of reptiles, and it is probable that it is in some essential features divergently specialized rather than primitive with respect to the placentals.

Discussion has been mostly in terms of single characters or structures, not animals as whole organisms. In some contemporaneous sequences, especially lines within single species or in chains of closely related species, the sequence as a whole may closely resemble the actual ancestral sequence (or chronocline). (That and other aspects of clines with respect to identification of primitive and advanced characters have been discussed at length by Maslin, 1952, and some of his conclusions are also applicable to sequences other than in low-level clines.) It is, however, the usual rule that in a given sequence no one taxon will be more primitive than any others in *all* respects. A tapir is more primitive than a horse in foot structure, but the tapir's nose is much more specialized. A platypus is usually cited as the most primitive of mammals because it lays eggs and has a reptilelike shoulder girdle and some other reptilian features. It is, nevertheless, highly specialized among mammals in its dentition, beak, amphibious adaptations, and some other particulars. Such crossing specializations, one group specialized in one respect and another related group in another respect, help in inference as to propinquity of descent and characteristics of the ancestry. The ancestry must have combined characters near those less specialized in the two groups in question and must have lived before their more specialized characters evolved.

In some recent groups, such as the Artiodactyla, no surviving form can be judged particularly primitive overall. Primitive artiodactyls have been entirely replaced and must in many respects have been quite unlike any recent descendant, as is known in this case to be true from fossil evidence. In other groups most of the steps in the ancestral-descendant sequence survive without profound modification. That is notably true among the primates. Tree shrews–lemurs–tarsiers–monkeys–apes–man are sequential in many characters, and it is reasonable to consider them sequential as whole organisms, despite many divergent specializations at lower levels within each step of the sequence. In characters that are sequential through the Primates, the tree shrews are consistently most primitive. The sequence among the recent primates and its bearing on primate phylogeny have been worked out well and in some detail by Le Gros Clark (1959).

NOTE ON PARALLELISM

As defined above, parallelism is the independent occurrence of similar changes in groups from a common ancestry and *because* they had a common ancestry. Some students (for example, Haas in Haas and Simpson, 1946) have preferred a more purely descriptive definition, especially by the geometrical model of parallel lines, symbolizing two lineages both changing but not becoming significantly either more or less similar. Pushed to an extreme that is logical but might be repudiated by proponents of the definition, parallelism would be said to occur between a lineage of oysters and one of mammals, both of which were becoming larger without other noteworthy change. That is not a particularly useful taxonomic concept. Most taxonomists do, however, consider that the term parallelism should be used only when community of ancestry is pertinent to the phenomenon.

Parallelism may be difficult or practically impossible to distinguish from homology on one hand and convergence on the other. The three

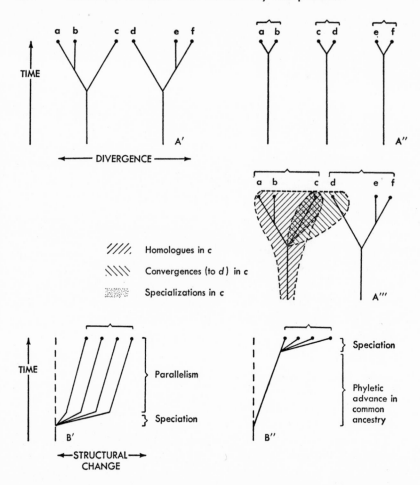

FIGURE 4. TAXONOMIC INTERPRETATION OF CONVERGENCE AND PARALLELISM

A. Diagram of six recent species, *a–f*, among which *a–c* are postulated as of one ancestral origin and *d–f* as of another; *c* and *d* are postulated as convergent toward each other. A'. Postulated true phylogeny. A". Placing of the species in three genera, symbolized by the braces, on the incorrect interpretation that the convergence in *c–d* is homology; the implied phylogeny shown is topologically quite different from that in A' and it is false. This presentation of the situation in both A' and A" is, however, so oversimplified as to be grossly unrealistic, the placing of *c* being based only on its characters convergent toward *d*. A'''. Still oversimplified but more realistic representation of the situation as regards *c*. In actuality *c* will certainly have characteristics: (1) linking it homologically with *a* and *b* and with the common ancestry of *a–c*; (2) linking it convergently with *d*;

processes, distinct enough in their characteristic manifestations, do intergrade through marginal cases. Fortunately, it is improbable that pure parallelism in a whole organism and at equal rates in all lineages ever occurs. It is always accompanied by homologies and perhaps also always, certainly usually, by divergent specialization in each lineage. It also commonly occurs at different rates in different lines. The lines may be sorted out on their divergent characters, and any convergence involved can usually be recognized by criteria already discussed. Remaining resemblances are then either homologous or parallel. The distinction between homology and parallelism can then frequently be made from the fact that homologous characters are primitive for the group but parallel characters are not. If, for instance, in a group of lineages with common ancestry one has low-crowned teeth and the others high-crowned, low-crowned teeth are primitive for the whole group, and evolution of high crowns in the other lineages was probably by parallelism. Michener (1949) has given a more complex and highly enlightening example of the sorting out of effects of parallelism in a large family of moths.

FIGURE 4. (*Continued*)

and (3) peculiar to *c* as specialization in its own lineage. Correct analysis of these characteristics gives the two genera represented by braces in A''', the phylogenetic implications of which are correct. As shown, the analysis of characteristics applies only to *c*. The other species will also have different groups of characteristics with varying taxonomic implications.

B. Diagram of four recent species postulated as having early diverged by speciation from a common ancestry and as having subsequently evolved in parallel. B'. Phylogeny postulated as correct. B''. Interpretation of recent species without recognition of the parallelism, the incorrect implication being that characteristics really parallel evolved in the common ancestry and that speciation occurred late. Although less misleading than in A, the representation in B is also grossly oversimplified by the omission of many characteristics with varying taxonomic value or implications. As between B' and B'', choice on data from the terminal species alone might be difficult, but sequential relationships would probably lead an experienced taxonomist to prefer the interpretation shown in B' and postulated as correct. (The broken vertical line projects the primitive condition for comparison with the recent species and makes their sequential relationship clearer.) In any event, a higher taxon uniting the four species, symbolized by the upper braces, would be phylogenetically valid and would be inferred from the incorrect as well as the correct interpretation. Furthermore, the pattern of B'' is topologically identical with B' and is phylogenetically incorrect only in a comparatively minor difference in time relationships.

Parallelism has several theoretical bases that help one to understand and also to recognize it. The structure of an ancestral group inevitably restricts the lines of possible evolutionary change. That simple fact greatly increases the probability that among the number of descendant lineages several or all will follow one line. That probability will be further reinforced by natural selection in a geographically expanding and actively speciating group if the ecologies of diverse lineages remain similar in respect to the adaptations involved in the parallelism. The degree of dependence on similar ecology resembles that of convergence, but the retention of homologous characters from the relatively near common ancestry usually distinguishes parallelism. The parallel lineages (unlike those only convergent) furthermore start out with closely similar coadapted genetic systems, and similar changes are more likely to keep the systems adequately coadapted. Tendency toward genetic parallelism is also strongly reinforced by recurrent "homologous" [12] mutations and similar relative mutation rates.

In the end, parallelism cannot always be distinguished from homology, but that usually does not matter very much. Like homology, parallelism does depend on community of ancestry. At worst, it may exaggerate the degree of propinquity of descent without falsifying its nature or the topological phylogenetic pattern. It may not even do that, because the mechanisms involved guarantee that on an average parallelism will tend to be closer in lineages closer to a common ancestry. Similarities developed by parallelism, whether recognized as such or not, do tend to be reliably proportional to propinquity of descent. The distinction of parallelism from convergence is vital, but is usually fairly easy if the convergence has not also been affected by community of ancestry (that is, does not really have an element of parallelism). The distinction of parallelism from homology is more difficult, but fortunately it is not so vital. That is not to say that parallelism is unimportant in classification; examples of its importance will be discussed in Chapter 6.

Correct and incorrect taxonomic interpretations of contemporaneous species with convergence and parallelism are shown diagramatically in Figure 4.

---

[12] This is a different usage of "homologous" from that usual in taxonomy, but the meaning is fairly obvious. Such mutations in parallel lineages are in fact also parallel, not homologous, in taxonomic terms.

4

From Taxonomy to Classification

Taxonomy is a science, but its application to classification involves a great deal of human contrivance and ingenuity, in short, of art. In this art there is leeway for personal taste, even foibles, but there are also canons that help to make some classifications better, more meaningful, more useful than others.

That the best basis for classification is evolutionary has already been decided. That means in the first place that a classification should be *consistent* with all that can be learned of the phylogeny of the group classified. That classification can or should express phylogeny is an evident error. Given consistency with its proper basis, classification should be as stable as possible, changed only when changing knowledge tends to make it definitely less useful or reveals inconsistency.

Discussions of the "reality" or "objectivity"—or "unreality" and "subjectivity"—of taxa tend to be rather futile or mere semantic squabbles. With special definitions, the concepts arbitrary and non-arbitrary do helpfully apply. Some procedures in classification must be arbitrary, which entails no drawback in principle but requires attention to procedures.

A basic canon of consistency with phylogeny involves monophyly, often defined vaguely or impractically. We must seek a better definition and assess its applications. We must also assess the interplay of different kinds of relationships in evolution, relationships of progression and stabilization, relationships vertical in ancestral-descendant lines and horizontal among contemporaneous groups, comparative weightings of divergence and diversity, and the related merits or demerits of defining categories more or less broadly. We must also consider the troublesome and frequently confused bearings of the relative antiquity of taxa on their relative ranks.

WHAT DO TAXONOMISTS DO?

Practicing taxonomy usually means constructing classifications of particular groups of organisms and also identifying specimens in accordance with a classification already made. Back of this practice and of still deeper importance are the theories, principles, rules, and procedures that are, strictly speaking, the science of taxonomy as here defined and that are the subject matter of this book. Formulating and testing those theories, principles, and so on are, of course, the most basic things a taxonomist does. Some of them have now been sketched. The next question is, by what taxonomic principles and procedures is an actual classification to be produced? When, as he certainly should, a taxonomist is going to practice classification, the technological application of his science, just how does he go about it? In general, he does the following things, more or less in this sequence but usually with a good deal of overlapping, jumping, and backtracking:

1. He selects or obtains the organisms to be classified. He may study them in the field and leave them there alive, but usually they are collected as museum specimens, preferably but not necessarily by the taxonomist who later classifies them. For large groups, at least, he will almost always need to use older collections, perhaps scattered in different institutions.

2. He observes and records data about the organisms, and also assembles from the literature as nearly as possible all the data already recorded about the particular groups he is studying. The needed data are of many different kinds, as discussed in Chapter 3. This activity overlaps the first, because some data, minimally those of locality and (for fossils) horizon, can only be observed in the field, and other, or in exceptional cases, all the data may be obtained in the field.

3. He sorts the organisms into taxonomic units, demes (local populations, more precisely defined in Chapter 5), subspecies, or species as may be appropriate. He then analyzes the data on polymorphic

forms and all kinds of variations. The analysis is statistical in principle and now is generally also so in details of practice, inferring population characteristics from the sample in hand. This actively overlaps the two preceding and also the two following activities.

4. He makes comparisons among the characteristics of the varying units, with special attention to the kinds and degrees of resemblances, differences, and sequences shown. These procedures, too, are largely statistical in principle.

5. He interprets the relationships revealed by comparisons in terms of basic taxonomic concepts, especially: homology, parallelism, convergence, primitiveness, specialization, as discussed in Chapter 3.

6. On the interpretations made in the last procedure, he bases inferences as to the evolutionary pattern among the populations studied. He may draw an actual diagram, preferably an inferred phylogenetic tree or a dendrogram intended to be topologically as nearly as possible equivalent to a phylogenetic tree. He may use some other visualization or may proceed in wholly verbal form. In any event, he reaches conclusions as to affinities such as are represented by a tree. His conclusions will also involve some considerations, as of degrees of divergence, that cannot readily be shown by a tree diagram.

7. He translates his conclusions on affinities, divergence, and so on into hierarchic terms, assembling and dividing the various groups of organisms into taxa of various ranks.

8. According to rules and usages, he selects from previous publications names applicable to the recognized taxa, and when there are no applicable published names he coins new ones.

Some of these procedures have already been discussed in the first three chapters. Others will be considered more fully in the following chapters. Some, such as methods of collecting specimens, although necessary if taxonomy is to be pursued, are not per se pertinent to our theme and so are omitted. A few, although more nearly pertinent, are nevertheless omitted on other grounds. That is especially true of the technical statistical procedures useful in steps 3 and 4, above. In detail, they are not essential to an understanding of taxonomic principle. Discussion of them adequate enough to be really useful would require much more space than is here available; and they have recently been adequately presented elsewhere (Simpson, Roe, and Lewontin, 1960).

In this chapter we shall be concerned, first, with some of the practical aims and requirements of classification not hitherto sufficiently discussed, and then with some special principles and procedures mostly involved in the seventh step of the preceding list.

CLASSIFICATION AS A USEFUL ART

Like many other sciences, taxonomy is really a combination of a science, most strictly speaking, and of an art. Its scientific side is concerned with reaching approximations, hopefully believed to be successively closer as the science progresses, toward understanding of relationships present in nature. One of the dictionary definitions of "art" is "human contrivance or ingenuity," and taxonomy becomes largely artistic, in that sense, when applied to construction of classifications. Nomenclature is completely an art and not a science at all, because it is solely a human contrivance and *corresponds with* nothing in (nonhuman) nature even though *applied to* scientific interpretation of things existing in nature. The recognition and arrangement of taxa at various levels has a scientific content, but it is also largely intermingled with art, requiring human contrivance and ingenuity. It has already been mentioned and will become more and more evident in this and the following chapters that even if interrelationships in a group of animals were completely known and even if there were complete agreement about the scientific principles to be applied, innumerable *different* classifications could be made consistent with those interrelationships and valid under those principles. Selection among those alternatives is decidedly an art.

A basic principle of taxonomic art is that its results should be useful. In classification this entails, among other things, three especially important subsidiary principles:

1. The basis of classification should be the most biologically significant relationships among organisms and should bring in as many of those as is practicable.
2. Classification should be consistent with the relationships used as its basis.

3. Classification should be as stable as it can be without contravening the two preceding principles.

The first point has already been sufficiently discussed. The second and third are especially pertinent here. The first and second limit the extent to which the third can be achieved in practice. Evolution, here adopted as the essential broad basis of classification, is constantly becoming better known both in detail and in breadth. Better grasp of its theories and principles must involve some changes in classifications in order to keep them most usefully related to biological science as a whole. Knowledge of particular groups of organisms is also still increasing greatly, for instance and strikingly at present, knowledge of their biochemistry. Even in the comparatively well-plowed field of anatomy new data are constantly added and many are still lacking. Such new data make possible firmer inferences and may contradict or overbalance older inferences previously involved in classifications. New groups of animals are still being discovered. The average rate of discovery is still accelerating in paleontology. In some recent groups (for example, birds) it is approaching zero. In others (for example, insects) it is still high, but for the most part in lower taxa. Recent high taxa (for example, those exemplified by *Neopilina* and *Hutchinsoniella*) do, however, continue to turn up. Not only must those new groups of organisms be inserted in classifications but also they usually demand lesser or greater changes in the framework of classification of previously known animals.

To maintain greatest usefulness, classification must be consistent not with knowledge of some fixed time in the past, but as nearly as may be with the constantly changing knowledge of today. It is therefore desirable that classifications should not remain static but should change continually as pertinent knowledge expands. Inevitably this desirable progress does entail some impediments to the usefulness of classifications. Classification is an absolutely essential means of conceptualization, communication, and storage of information about animals. In terms of information theory, a great deal of noise is introduced when classifications (and their nomenclatures) vary from time to time and from person to person. In this respect, the stablest classification is certainly the most useful. Although of no great importance for zoology as a whole, it is also of enormous practical significance to custodians of collections that changes in classifications can involve a heavy burden of otherwise unfruitful labor in changing cataloguing, indexing, and storage systems based on them.

There must be some compromise between the usefulness of up-to-date classifications and the usefulness of stable classifications.[1] The following compromise is certainly the best and deserves strong emphasis:

A published classification in current use should be changed when it is definitely inconsistent with known facts and accepted principles, but only as far as necessary to bring it into consistency.

Taxonomists, most of whom are strong individualists, frequently do not follow that rule in practice, but their failure to do so is a real disservice to others. It is true that application of the rule is not automatically clear in particular instances. There are often different classifications in current use for the same group. Successive rules to apply in that case are:

1. If one classification is more consistent than others, adopt it with, if necessary, such modification as will make it as fully consistent as possible.

2. If there is no evident choice as to consistency, adopt and if necessary modify the classification in widest use.

3. If there is no evident weight of usage, adopt and if necessary modify the most authoritative classification, giving due weight to the experience of the classifier, the soundness and multiplicity of the data used by him, and the acceptability of his taxonomic principles and procedures.

The concept of consistency is essential in these rules, but it has not yet been defined here. That is a complicated matter that requires consideration at some length. It has often been and still is occasionally said that the purpose of evolutionary classification is to express phylogeny. It is, however, true that no form of classification yet devised, certainly not the Linnaean hierarchy, is really able to express phy-

---

[1] It is true that many a museum curator has had to eschew compromise and choose stable, i.e., out-of-date, classification in the physical care of his collections from sheer lack of staff time. He need not and should not do so, however, for purposes of study and communication.

logeny, at least not in the sense of presenting it fully and unequivocally. That was already clear to the founder of evolutionary taxonomy. Referring to his diagram of phylogeny, Darwin (1859) wrote:

If a branching diagram had not been used, and only the names of the groups had been written in a linear series [as they are in a hierarchic classification], it would have been still less possible to have given a natural arrangement; and it is notoriously not possible to represent in a series, on a flat surface, the affinities which we discover in nature among the beings of the same group.

Evolutionary taxonomists have always recognized the fact already considered notorious by Darwin, but they have sometimes said that classification expressed phylogeny when they meant that it is *based on* phylogeny or is a *partial* expression of it. Critics of evolutionary classification had no difficulty in demonstrating that it is inevitably an inadequate expression of phylogeny, but when they concluded that it is therefore impracticable on its own premises they were mistaking the premises and attacking a straw man.

The statement that evolutionary classification is based on phylogeny has also been open to misunderstanding. It has been taken to mean that such classification follows entirely from lines of descent and their branching as shown in the usual diagram of a phylogenetic tree. That (as Darwin also already knew) is still not entirely true. Evolutionary classification commonly requires information that is still largely phylogenetic information but that is practically impossible to include in a tree diagram. It also necessarily requires some arbitrary divisions that can be *applied to* the tree as a basis but that do not *arise from* the tree as originally drawn. These facts, while they do not really make a classification any less phylogenetic in basis, have made the term somewhat equivocal, and I therefore prefer the designation "evolutionary" to "phylogenetic" for the kind of classifications I am talking about.

It is preferable to consider evolutionary classification not as expressing phylogeny, not even as based on it (although in a sufficiently broad sense that is true), but as *consistent* with it. *A consistent evolutionary classification is one whose implications, drawn according to stated criteria of such classification, do not contradict the classifier's views as to the phylogeny of the group.* We must, then, next discuss the criteria that are used to draw up an evolutionary classification and see what their implications are. The principal kinds of criteria to be considered are as follows:

1. Criteria related to objectivity, reality, arbitrariness, and the like.
2. Criteria related to monophyly, polyphyly, clades and grades.
3. Criteria related to the different kinds and degrees of affinity involved in phylogeny.
4. Criteria related to the relative antiquity of taxa.

There is so much to say about each of these that they will be discussed seriatim in following sections of this chapter.

OBJECTIVITY AND ARBITRARINESS

Much paper has been used, and in my opinion most of it wasted, on arguments as to whether taxa are or can be "real" or "objective." The discussion was already raging in the time of Linnaeus, who concluded essentially (but in different terms) that species may be and genera usually are objective but that higher categories are not. The most frequent but far from undisputed conclusion nowadays is that species, when correctly defined, are real or objective and that other categories are not. (See references in Simpson, 1951; see also the discussion of species in Chapter 5 of this book.) The discussion and that questionable conclusion do indeed have a point, but it has usually been put in the wrong or at least in confusing terms. Any definition of a taxon, whatever its basis as long as taxonomically relevant,[2] distinguishes a set of real, objective organisms. On the other hand, the concept of a taxon, the thing really present in the classifier's mind and named and referred to by him, is invariably subjective, whether for a species or any other category. So are the definitions proposed and the names applied: they have no real relationship to any objects in nature except through the subjective processes of a taxonomist. Thus any taxon whatever is in some respects or in some application of the words real or objective and in others simultaneously unreal or subjective.

[2] Examples of obviously irrelevant definitions: "insects with eight pairs of legs," "animals with one or more nuclei." The first applies to no animals, the second to all animals. Neither one really defines anything.

The point properly at issue in such discussions as have not been entirely confused or purely semasiological is whether the groups distinguished as taxa are continuous or discontinuous in biologically significant, defined, and determinable ways. In order to clarify that point, I have elsewhere suggested (Simpson, 1951) that the terms nonarbitrary and arbitrary be used instead of real and unreal, objective and subjective. A group is nonarbitrary as to inclusion if all its members are continuous by an appropriate criterion, and nonarbitrary as to exclusion if it is discontinuous from any other group by the same criterion. It is arbitrary as to inclusion if it has internal discontinuities and as to exclusion if it has an external continuity. For species as usually defined among recent animals (see Chapter 5), the appropriate criterion of continuity is potentiality for frequent and continuing interbreeding, sequential both vertically in time and laterally through the group. Thus the point made by those who consider the species as objective and other taxa as subjective can be more clearly stated thus: the species (genetically defined) is the one taxon that is usually nonarbitrary both as to inclusion and as to exclusion. Infraspecific taxa are usually nonarbitrary as to inclusion but arbitrary as to exclusion. Supraspecific taxa are usually nonarbitrary as to exclusion but arbitrary as to inclusion. (See Figure 5A.)

In spite of various complications and problems, some of the more important of which will be dealt with in Chapter 5, that criterion is usually adequate in itself to distinguish evolutionary infraspecific, specific, and supraspecific taxa *in contemporaneous animals*. Even among them, however, question arises as to what criteria may be used for the arbitrary division of species into infraspecific taxa and the arbitrary union of species into supraspecific taxa. The question of defining the categories to which such taxa belong must be deferred until we have more kinds of criteria in hand. Two possible relative criteria that are logically related to arbitrariness may, however, be mentioned here.

Within a species there may be lines, usually traceable in geographic terms, across which there is less interbreeding than within populations on each side of the line. Division of the species into infraspecific taxa, such as subspecies, along those lines is less arbitrary than if it cut through a more freely interbreeding population. A subspecies may in such instances, which are far from universal, be comparatively, but not absolutely, nonarbitrary both in exclusion and inclusion. (Figure 5B.)

Among a number of species some will resemble each other more than others. It is to be assumed that the resemblances, as far as homol-

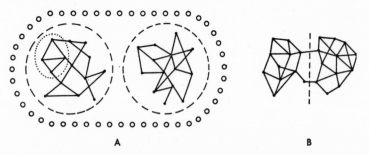

A   B

FIGURE 5. ARBITRARY AND NONARBITRARY ASPECTS OF CONTEMPORANEOUS
TAXA WITH RESPECT TO GENETIC CONTINUITY

A. General relationship of arbitrariness to infra– and supraspecific taxa. Individuals are represented by heavy dots. Each of the two species shown, enclosed in broken lines, is a continuous nexus of sequentially interbreeding individuals. Each is nonarbitrary as to inclusion and exclusion. In one a subspecies, enclosed in a dotted line, is shown, and it is seen to be non-arbitrary as to inclusion but arbitrary as to exclusion. The two species are united in a genus, enclosed in a line of open circles, which is nonarbitrary as to exclusion but arbitrary as to inclusion.

B. Diagram of a genetic nexus of a species divided into subspecies along the broken line, with less interbreeding between than within the two subspecies.

ogous, reflect propinquity of descent, and that brings in other kinds of criteria to be discussed later. Those other criteria should also be applied in each case, but a criterion related to arbitrariness may be derived from the resemblances themselves. If there are groups of species such that all the species in a group more closely resemble some other in that group than they do any species in any other group, then equation of the groups with supraspecific taxa is comparatively nonarbitrary. (Figure 6.) (The criterion for comparative continuity implied is of course degree of resemblance.) Mayr, Linsley, and Usinger (1953) have defined the categories genus and family mainly on this basis, by the presence of "a decided gap" between the corresponding taxa. Application of the criterion is often possible and is desirable when possible. It is not, however, universally possible and is still less universally applied. It cannot, therefore, be used as the sole criterion for defining those categories.

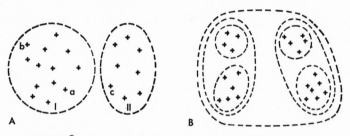

FIGURE 6. GROUPING OF TAXA BY GAPS IN RESEMBLANCE

Species are symbolized by crosses, higher taxa by broken lines. Spatial separation symbolizes degree of resemblance.

A. A comparatively nonarbitrary arrangement of groups of species. Species *a* in Group I is actually farther from some species, e.g., *b*, in the same group than from some species, e.g., *c*, in Group II. Nevertheless, *a* is sequentially linked to *b* by gaps all smaller than that between *a* and *c*.

B. A more clear-cut grouping of species into taxa of three successively higher categories.

When the time dimension is introduced, application of a criterion of arbitrariness becomes more complex or largely impractical. All animal taxa are genetically continuous with some others if they are followed back in time. Separate branches are discontinuous after they have arisen, and so the criteria of arbitrariness or nonarbitrariness are applicable to them. When the whole pattern through time is considered, any taxa within the pattern may still be made nonarbitrary as to inclusion: that involves the criterion of monophyly, next to be considered. In the larger pattern, however, no taxa (including species) can be strictly nonarbitrary as to exclusion. One must somewhere draw a completely arbitrary line, representing a point in time, across some steadily evolving lineage and say, "Here one taxon ends and another begins." In principle, that means that an individual animal could belong to one species one instant and to another species the next instant. That is seemingly absurd, but no more so than results of other available criteria for such necessary arbitrary divisions. In actuality, our data are not yet so complete that these absurd cases arise in practice.

There are phenomena that mitigate the arbitrariness of this procedure. A few students (for example, Schindewolf, 1950; Goldschmidt, 1940) maintain that species and higher categories normally arise abruptly, in one sharp step. If so, each such step would make a nonarbitrary division point. The phenomenon can occur, for example, as a consequence of polyploidy, but the consensus (with which I heartily concur) is that it is unusual, especially in animals, at any level and that it does not mediate changes from one high level taxon to another. It does, however, seem to be common without being universal that changes initiating what will later become a new taxon, and especially at higher levels, occur at exceptionally rapid evolutionary rates by what I have called quantum evolution (Simpson, 1944, 1953). Then the *period* of rapid transition may serve for comparatively nonarbitrary division, but the exact *point* of separation remains as arbitrary as ever.

Only paleontologists have to deal with this problem in practice, and in this respect their data have defects that are often practical virtues. All temporal sequences of known, collected fossils have gaps, time intervals in which fossils of the particular group are unknown, if they are extended over long periods. Although these are only discontinuities in the data, not in the actual phylogenies as they were lived, they do furnish division points that are nonarbitrary as far as the materials actually being classified are concerned. All paleontologists do use them as such division points. Such gaps also have the practical advantage that they tend to be more striking and more consistently present for higher than for lower taxa. (That fact has been adduced as evidence for actual discontinuities in evolution, but it is readily explicable as a consequence of sampling phylogenies that were really continuous; see Simpson, 1953, 1960a.)

The necessity for making arbitrary divisions in continuous temporal sequences still remains and is of interest, both in the theory of taxonomy and in the practice of classification. Many short or moderately long essentially continuous sequences of ancestral-descendant species and genera are now known, and they demand arbitrary division into taxa. New discoveries constantly tend to fill in gaps at all levels. For example, the transition from class Reptilia to class Mammalia is becoming fairly well represented by known fossils, and the question of where to draw the arbitrary line between the classes is increasingly disputed. (See three notes by Van Valen, Reed, and Simpson, Evolution, in press.) It is, of course, a question more of taste or skill in the art of classification than of theory in the science of taxonomy.

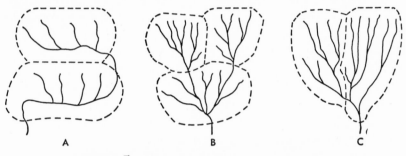

FIGURE 7. THREE HYPOTHETICAL PHYLOGENIES

These are analogous to many that really occur; they are divided into taxa as shown by the broken lines. A is evidently less arbitrary, in a broad and somewhat ill-defined sense, than B, and B somewhat less than C.

Even with the time dimension present and on the assumption that phylogenies are adequately known, certain groupings may appear definitely less arbitrary than others, using "arbitrary" in a broader, less technical sense than in the previous definition. The point is adequately made by Figure 7, and is also demonstrated by concrete examples in Chapter 6. Other considerations, as of monophyly and degrees of divergence and diversity, enter into such cases.

When there is a choice between the two, a nonarbitrary arrangement will usually be preferred to one that is arbitrary. It is, however, virtually impossible to be completely nonarbitrary for any taxa other than species, and frequently also for species. There is nothing wrong with being arbitrary in the practice of an art, including the art of classification. An arbitrary element is indeed absolutely necessary, and groups arbitrarily delineated are no less "real" on that account. It is the division lines between them and not the groups themselves that are nonobjective.

MONOPHYLY AND POLYPHYLY

In broad terms, as in many dictionaries (with variations), monophyly is defined as evolutionary origin from one ancestry and polyphyly as derivation from two or more ancestries. Those and similar definitions are highly ambiguous and are also nonoperational, that is, they are so vague that they provide insufficient criteria for separating one from the other by analysis of evidence. If the concepts are to be used, and they must be, they must be given better definitions. That will shortly be attempted, but first consideration must be given to the fact that there are two related but quite distinct questions in this field, and that these are sometimes confused in the large literature of the subject (see, for example, Remane, 1956, and references there). The first question is about the extent and nature of polyphyly in the course of evolution. The second is whether and in what sense monophyly may be used as a criterion (a canon of the art) in the practice of classification.

It now seems curious that some early evolutionary taxonomists, notably T. H. Huxley (1880), concluded that the "laws of evolution" demand a high degree of polyphyly, a conclusion not supported by Darwin. Huxley thought that each order of mammals had separately passed through "hypotherian," "prototherian," "metatherian," and "eutherian" stages. The living monotremes, order Monotremata, have not yet evolved beyond the prototherian stage or the living marsupials, order Marsupialia, beyond the metatherian, but these orders were supposed to be entirely distinct from and not ancestral to, for example, the prototherian and metatherian ancestors of the now eutherian primates.[3] Even more extreme views have been held by a few later taxonomists, Kleinschmidt (1926), for instance, still supported an early post-Darwinian suggestion that each *species* of mammals had a separate *protozoan* ancestry, and Rosa (1931) maintained that any possible common ancestor of marsupials and placentals must have been an invertebrate.

Such views are so completely opposed by conclusive evidence that

[3] For Huxley, the "Hypotheria" were hypothetical, as the name implies, but his structural definition of them quite nearly corresponds with probable conditions among therapsid reptiles. Thus Huxley's opinion demands that each order of modern mammals have a separate reptilian ancestry.

we can only smile at them and pass on. That does not, however, dispose of the question at a different level and with more rational relationship to the evidence. Parallelism is a widespread phenomenon in evolution, and it is not uncommon to find that some generally recognized taxon arose by parallel evolution through two or more lineages from different ancestral taxa. A few examples at different levels are:

The extinct genus of horses (Equidae) *Merychippus,* as hitherto usually delimited, probably arose from more than one species of the genus *Parahippus,* as usually conceived (for example, Quinn, 1955).

The primate suborder Anthropoidea of most current classifications evidently arose from two (and possibly from three) lower taxa of the suborder Prosimii (for example, Le Gros Clark, 1959).

The classical order Edentata, as almost universally recognized from the late eighteenth century into the twentieth and still retained by a few authorities today, certainly had three quite distinct origins, probably from at least two different orders (for example, Simpson, 1945).

The class Mammalia, of virtually all classifications since Linnaeus, almost certainly arose from several different taxa within the order Therapsida, formerly always and usually now defined as part of the class Reptilia (for example, Simpson, 1959b).

Such examples—and they could be multiplied extensively—lead to the second question, how they are to be dealt with in the applied art of taxonomy. There are in general four possible approaches:

1. Revert to typological or to empirical taxonomy in which no phylogenetic criteria are involved.

2. In place of or in addition to a criterion of monophyly in evolutionary classification, accept criteria related to the attainment of grades of progression, which are after all just as evolutionary as monophyly and are also related to phylogeny in a less strict way.

3. Frame a more definite but still evolutionarily sound conception of monophyly that would make it relative to the ranks of the taxa involved and that would in some instances, not in all, make taxa that arise from more than one lineage still monophyletic by definition.

4. Accept an absolute definition of monophyly, tied definitely to some one taxonomic rank. There would then be two ways in which this more rigid criterion could be met:

a. The taxon formerly considered single could be split into two or more taxa of the same rank, each monophyletic by the definition adopted.

b. The taxon in question could be expanded backward in time to include ancestral groups, previously placed in other taxa, until an ancestral stem unified enough to meet the adopted definition of monophyly became included in the taxon in question.

Each of these procedures has been applied in particular classifications, and each has been supported as the sole ruling criterion by one taxonomist or another. For reasons that must already be fully evident, I feel strongly that the first should be adopted only *faute de mieux*. There are indeed groups in which a better solution is lacking at present, notably among protists but probably also among some metazoans about which we simply do not have enough information. For most metazoans, however, information either in hand or now practicably obtainable does permit one or more of the three evolutionary approaches. Those approaches, 2–4 of my list, are not mutually exclusive. Applications of any two or of all three of them will often give the same result, depending for the most part on just how monophyly is defined. All three seem to me to be valid in principle, that is, to be consistent with acceptable evolutionary taxonomic theory. Choice among them is not a matter of right and wrong but of artistic judgment in each individual case, into which the following considerations might enter:

Desirable coincidence of the results of applying two or all of the three approaches to the same case.

Practical usefulness in making taxa easily definable and recognizable while still evolutionarily valid.

Concordance with other canons of the art, such as those of divergence and diversity discussed later in this chapter.

Conformance with the rule that current classifications should not be changed unless that is really necessary for consistency with principle.

It has often been stated that approach 4a, with a very narrow and rigid definition of monophyly, is demanded by evolutionary classification. Resulting instability and inconvenience, demonstrable in many cases, have been given as arguments against evolutionary classification. It simply is not true, however, that evolutionary classification logically has any such requirement. Approaches 2, 3, and 4b are also in full logical accord with evolutionary taxonomy.

It is in fact almost if not quite impossible to frame a definition of monophyly that is both absolute and practical. If "mono-" were taken in the sense of "minimal," the appropriate definition would be descent from a single individual or one mating pair. But under such a definition of monophyly exceedingly few animal taxa could possibly be monophyletic. An apparently better definition, and one usual when the word is defined at all (most taxonomists fail to define it), would be descent from a single species. That can, nevertheless, readily be shown to be undesirable in principle and usually inapplicable in practice. To be actually definitive, the definition must imply that the single ancestral species is included in the taxon made monophyletic by its ancestral status.[4] Now, a great majority of genera have at one time or other had two or more species not all directly ancestral-descendant among themselves. If, then, just the one ancestral species is placed in the genus to which it gave rise, either (a) that one species is included in two genera, that in which it arose and that to which it gave rise, which violates the hierarchic system, or (b) it will usually be classed in the same genus with species to which it is less closely related than to species in the ancestral genus (for reasons analogous to the fact that a man is less closely related to his grandson than to his brother). The latter objection is not too serious in principle, because similar anomalies must occur whenever lineages are subdivided temporally, but it would frequently make insoluble problems in practice. The lateral affinities among contemporaneous species of a genus are generally more obvious than those of the whole genus to the one species ancestral to it, and they are much more often the basis on which a genus is and must be recognized. Even more crippling to practice under such a definition of monophyly is the fact that the one ancestral species is rarely actually known or identifiable as such, and that without identifying it the applicability of the definition is practically un-

[4] Otherwise the definition would make any aggregation of animals, such as a genus composed of what we now call *Homo sapiens* and what we now call *Drosophila pseudoobscura*, monophyletic, because it is probable that a single species ancestral to both did occur in *some* taxon.

ascertainable. Similar objections apply a fortiori to any criterion that taxa higher than the genus should be descended each from a single species included in them.

These and other considerations persuade me that definition of monophyly as descent from a single species or any other single taxon at one rank fixed by the definition is impractical in classification and unnecessary for consistency with taxonomic principles. The following definition is preferred:

Monophyly is the derivation of a taxon through one or more lineages (temporal successions of ancestral-descendant populations) *from one immediately ancestral taxon of the same or lower rank.*[5]

There are different degrees or levels of monophyly under this definition, and as a rule the level should be specified or evident in each case. The level of monophyly is specified by the category of the lowest ranking single taxon immediately ancestral to the taxon in question. The level of *polyphyly* is specified by the category of the highest ranking taxa two or more of which were immediately ancestral to the taxon in question. Thus in the examples previously given, *Merychippus* is considered monophyletic at the generic but polyphyletic at the specific level.

Subspecies or species are defined on other criteria that override any criterion of monophyly. Subspecies may arise from other subspecies, but doubtless also arise from parts of populations not given taxon status in classification. Species are usually minimally monophyletic, that is, at the specific level, but they may be polyphyletic without thereby violating current taxonomic theory or practice, that is, be partly or fully interspecific hybrid in origin. Above the species I suggest the following criteria:

All supraspecific taxa should be minimally monophyletic as far as the evidence at hand indicates, that is, a T_j taxon should be monophyletic at the C_j level, a genus of organisms at the generic level, a family of organisms at the family level, etc.

[5] Beckner (1959) has followed the same line of thought to a definition that is almost identical in concept:

"A taxon of any rank will be said to be monophyletic at the level of taxa of rank j whenever each member of the taxon is either a first-generation descendant of members of that taxon, or a first-generation descendant of members of a distinct taxon of rank j."

It will be remembered that by set theory the members of taxa are individuals, not other taxa. The members of a genus are organisms, not species.

Other things being equal (that is, if other criteria do not definitely outweigh this preference), a supraspecific taxon T_j should be monophyletic at the C_{j-1} level, or lower if the evidence and the balance with other criteria permit.

Of the examples earlier given, the Anthropoidea and Mammalia meet the second, more stringent, of these criteria. *Merychippus* meets the first, less stringent criterion but not the second. Whether in this case other criteria be considered as overbalancing the more stringent criterion is a matter for decision by the individual classifier. In my opinion the stringent criterion is frequently overridden by other criteria for genera, rarely for higher taxa. The Edentata as originally defined are not monophyletic by either criterion and have in fact been broken up into three separate monophyletic orders by almost all recent taxonomists. This definition of monophyly and the criteria based on it are both theoretically justified and practically useful. They are strictly phylogenetic in basis. They assure that the members of a taxon meeting the criterion will be specially related and that the degrees of affinity will tend to be adequately correlated with the rank of the taxon. Information on hand or available for most, at least, of the groups of metazoans suffices to determine with adequate probability whether the appropriate criterion is met. (As usual, *certainty* is impossible and cannot be insisted upon.) These concepts of monophyly and polyphyly are, furthermore, completely unambiguous when the levels involved are specified, and any such concept that does not specify levels is *ipso facto* ambiguous.

GRADES AND CLADES

A little more attention will now be given to a matter touched in passing while discussing monophyly: groups of animals similar in general levels of organization as distinct from groups of common genetic origin. J. Huxley (1958) has recently discussed this subject in a lucid and useful way and has proposed that the former groups be called *grades,* the latter, *clades.* The distinction was already being made in other terms by the earliest evolutionary taxonomists, notably by T. H. Huxley in connection with his views on mammalian classification mentioned earlier in this chapter.

It is a frequent phenomenon in evolution that whole groups of animals, with numerous separate lineages, tend to progress with considerable parallelism through a sequence of adaptive zones or of increasingly effective organizational levels. (On the former, especially, see Simpson, 1953, where many examples are given and the whole phenomenon is considered at some length; on the latter, especially, see Huxley, 1958, who stresses this aspect of grades.) In some instances the progression is fairly steady, as in that of bony fishes toward the teleost level in the Mesozoic. In others there is comparatively rapid change from one level to another with subsequent stabilization and lesser diversification at each level after it is attained. Then there are more or less clearly definable steps, as between browsing and grazing horses or between prosimians, monkeys, apes, and men. It is to such steps that the concept of a grade most clearly applies.

The existence of a gradelike progression may not be obvious from the lineage pattern of phylogeny, alone, but it may also be clearly reflected in that pattern. (See Figure 8.) In either case grades have a bearing on classification and entail certain problems. The following diagram, modified from Bather (1927), presents one of these in radically simplified form:

First	Second	Third	
a	*b*	*c*	Italic
a	b	c	Roman
α	β	γ	Greek

The letters may be taken as symbolizing species, "First," "Second," and "Third" lineages, and "Italic," "Roman," and "Greek" grades. As Bather emphasized (and see also Pirie, 1952), for practical purposes it does not matter at all whether we call *a* "First Italic" or "Italic First," that is, whether we have a lineage genus "First," with species *a*, a, and α, or a grade genus "Italic" with species *a*, *b*, and *c*. In fact both approaches have frequently been followed, among graptolites, for example, in which the genus *Monograptus* originally designated a grade and is still sometimes used in that sense. That grade is, however, now known to have been reached independently by a considerable number of different clades to each of which a generic name is given in most recent work (for example, Bulman, 1955).

I would say that the choice between lineage genera and grade genera does not matter much either, provided that we can assume some relationships always present in nature but omitted from the oversimplified example, and further provided that the alternatives are really as simple and as evenly balanced as the example would suggest, which is almost unknown in actual practice. We must assume that these three lineages

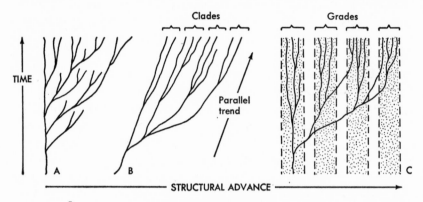

FIGURE 8. SOME DIAGRAMMATIC RELATIONSHIPS BETWEEN STRUCTURAL ADVANCE AND PHYLOGENY

A. With no clearly distinct grades and no taxonomically useful relationship between grades and clades; a common relationship in many animal groups.

B. With no clearly distinct grades but with strong parallelism in clades, each of which tends separately in the same direction up a continuous grade scale; similar but of course more complex patterns occur in bony fishes and in rodents, for example.

C. With grades as defined steps and clades more or less but not exactly corresponding with the grades; a similar pattern occurs in primates, as well as in other groups.

are closely related among themselves and that no one of them is more closely related to some other lineage than it is to the other two shown. If that is not true, then the grade arrangement is flatly invalid as evolutionary classification. In fact it would then rarely if ever be adopted even in supposedly nonevolutionary classification because the lineages would have consistent dissimilarities and would not be so simply parallel. Thus the lineage genera are acceptable in any case, but the grade arrangement is acceptable only if it does in fact also meet a phylogenetic criterion, although not the extreme criterion of specific monophyly. The grade genera are monophyletic at the generic level, hence minimally meet our previous criterion. Moreover the grades are evolutionary, and if the minimum requirements of monophyly are met this is just as evolutionary a procedure as the use of lineage genera.

In real phylogenies, with their complex branching and their interweaving of homology, parallelism, divergence, and convergence (not parallelism alone, as in Bather's treatment), grades and clades are rarely, I think never, so perfectly balanced at right angles to each other. Usually, indeed, they run parallel, and grades and clades have a strong tendency to coincide, as in Figure 8C. The concept of a grade has little meaning unless it is applied to related animals only. Both insects and fishes have (convergent, not parallel or homologous) eyes, brains, jointed limbs, etc., but it would obviously not be a useful taxonomic concept to place them therefore in the same grade. At a minimum, grades must be based on parallelism, not convergence, and their use certainly cannot eliminate a phylogenetic basis.

A clade, by definition, is monophyletic and the possible coincidence of grades and clades will in most cases depend on the level of monophyly adopted. The examples of grades in horses and primates given above meet my criterion for level of monophyly in classification. So do Huxley's examples and indeed all others I can think of, although exceptions probably do exist. (This is in part an artifact, because in possible examples that do not meet the criterion, as in an order Edentata including pangolins, aardvarks, and armadillos, the conclusion is that the supposed grade was not meaningfully defined!) If a serious issue arises, I would tend to prefer clade taxa, that is, to use as low a level of monophyly as practicable. If, however, some classifier makes grade taxa, either as a matter of his preference in the art or though mistaking parallelism for homology, I consider those taxa valid evolutionary groups—always supposing that the evidence does not oppose their meeting the minimal requirement of monophyly by the criterion I accept.

Huxley has tentatively proposed that we might use two separate classifications and nomenclatures, one with grade taxa and one with clade taxa. The concepts and terms grade and clade are extremely useful in understanding and discussing evolutionary phenomena and classification. I cannot, however, see any necessity or particular usefulness in having two systems of classification on that basis.

Is a man more closely related to his father, son, or brother? The actual genes involved may be quite different, but the degree of genetical relationship to father and to son is invariably the same (0.5 in terms of proportion of shared chromosomes). Genetical relationship to a brother is variable—from 1.0 to 0.0 in terms of chromosomes, although the probability of those extremes is exceedingly low—but the mean value is the same as for father or son. Unfortunately, relationships among taxa do not have such fixed a priori expectations, and they cannot be precisely measured. The same two kinds of relationships nevertheless exist: among successive taxa in an ancestral-descendant lineage, and among contemporaneous taxa of more or less distant common origin. In accordance with the usual coordinates of tree representation, the former relationships are called *vertical* and the latter *horizontal*.

One kind of relationship is obviously just as objective as the other and may be just as close or distant. Moreover it seems obvious that the two are equally phylogenetic: in the human analogy it would plainly be ridiculous to say that the relationship of brothers has nothing to do with their ancestry, or phylogenetic origin. We therefore need to waste no time on the manifestly erroneous idea, too often expressed both by friends and foes of evolutionary classification, that only vertical relationships are phylogenetic. (Osborn, an extreme exponent of vertical classification, was wont to say that only vertical relationships are evolutionary!)

In dealing with recent animals or with contemporaneous faunas of fossils, only horizontal relationships are *directly* involved. They are, however, interpreted in the light of their implications as to propinquity of descent, and conclusions about vertical relationships do thus enter into classifications of contemporaneous animals. In temporal sequences of fossils vertical relationships are directly presented. In either case, the general principles of classification must refer to both

kinds of relationships, as more or less closely reflected in a phylogenetic tree or a topologically equivalent dendrogram.

Examination of any extensive tree or dendrogram at once reveals
that classification by either vertical or horizontal relationships alone
is absolutely impossible. (Incidentally, it is no less impossible for
empirical or typological classification, because the two kinds of relationships give different patterns of degrees of resemblance or different archetypes, which sooner or later conflict and demand switching
from one to another.) In translating the phylogeny into taxa a compromise must somewhere be effected; some divisions among taxa
must be horizontal and some vertical. Choice as to just how to effect
the compromise is part of the art of taxonomy, a matter of taste and
ingenuity. That is one of the two main reasons why one, agreed phylogeny is consistent with many different classifications. The other main
reason, splitting versus lumping, will be discussed later.

In my opinion, there are no good direct criteria for choice between
vertical and horizontal classification on this basis alone. There are
temperamental patterns, or one might say schools of art, among classifiers. Some (for example, Osborn), who are also likely to insist on
extremely low-level monophyly and to be extreme splitters, give top
priority to vertical relationships and compromise with horizontal relationships only when absolutely forced. Others (for example, Frech-
kop), who are likely to have a minimum requirement of monophyly
or none at all and to be extreme lumpers, direct attention to horizontal
relationships and minimize vertical relationships as much as possible.
Both extremes seem to me to be unjustified, as extremes usually are.
There is, however, plenty of room for proper artistry in the area of
moderation between the two extremes. Other criteria should be
brought in. Extreme absolute monophyly overemphasizes the vertical,
but the relative criteria of monophyly previously advanced tend to
preserve a vertical element without conflicting with the necessity for
horizontal divisions as well. The criterion of stability tends to favor
the horizontal element, because horizontal relationships are as a rule
easier to recognize (especially among recent animals), frequently
have priority of publication, and are less disturbed by new discoveries,
especially of fossils.

The necessity for compromise and its nature can be made sufficiently evident by one example (Figure 9). The mammalian order Carnivora appears in the Paleocene and Eocene with a number of groups, usually hierarchized as families, that have clearly diverged from a common ancestry but that are already sharply distinct when first known by fossils. One group, Miacidae, is related vertically to the now surviving suborder, Fissipeda, of terrestrial carnivores and is near to or includes its ancestry. As of the Paleocene and Eocene, however, the Miacidae have at least equally close horizontal relationships with

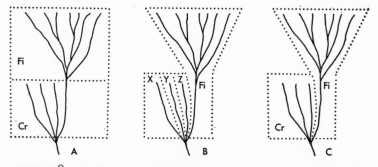

FIGURE 9. DIAGRAM OF HORIZONTAL, VERTICAL, AND COMPROMISE
CLASSIFICATION

The pattern is similar to the phylogeny of families of terrestrial Carnivora, with Miacidae the lower stem to the right.

A. Strictly horizontal major groups, as in Matthew's suborders Fissipeda (Fi) and Creodonta (Cr) of Carnivora.

B. Strictly vertical major groups, with three suborders (X, Y, Z) coordinate with Fissipeda.

C. Compromise, as in the usual current classification of Carnivora.

families without special vertical relationship to the Fissipeda. A now classical arrangement by Matthew (1909) puts all the Paleocene-Eocene groups, including Miacidae, and their later descendants except the Fissipeda in a suborder Creodonta. That is extreme horizontal classification of the families, but makes the necessary compromise by vertical classification within families. Vertical classification of families places the Miacidae in the Fissipeda, and a logical extreme would then be to place each of the other Paleocene-Eocene families in a separate suborder, because they are, on available evidence, as distinct from each other as from the Miacidae and of about equal antiquity.

That transfers the necessary horizontal division to the ancestry of all the "suborders" so created, which happens to be a convenient place for it at present because there are pronounced gaps in knowledge at the base of each of the postulated "suborders." Nevertheless, as far as I know no one has ever proposed that arrangement, evidently because it produces suborders extremely unbalanced in divergence and diversity (criteria next to be discussed). A compromise between the two extremes, which goes back to long before Matthew (in essentials but not in precise form to Cope, 1875), puts the Miacidae vertically in the Fissipeda but retains all the other Paleocene-Eocene carnivores in one horizontal suborder Creodonta. That has been a comparatively stable arrangement and is now general. (For fuller discussion, see Simpson, 1945.)

There is some tendency for categories to differ with respect to the balance of horizontality and verticality in their member taxa. That will be mentioned again in the chapters on lower and higher taxa. Here it may just be noted that, with much variation and some inconsistency: genera may tend to be horizontal bundles of species, especially in neontology; subfamilies and families may tend to have predominantly vertical relationships, especially in paleontology; and suprafamilial taxa tend toward a more even balance of the two.

DIVERGENCE AND DIVERSITY

It is clear enough that the conceptual and verbal usefulness of classification goes back to its most primitive basis in degrees of similarity and dissimilarity. As has been demonstrated, that is a necessary but not a sufficient basis. Similarities and dissimilarities differ in kind among themselves; they are often in conflict; they require criteria of ranking not provided by apriorism or empiricism; and without such criteria they are lacking in explanatory content and broader scientific meaning and usefulness. Those insufficiencies are made good by the principles and criteria peculiar to evolutionary classification. That does not mean that similarities and dissimilarities, per se, are no longer among the criteria useful in evolutionary classification. Consistency with phylogeny is a prime requirement of evolutionary classification, but that requirement can be met by numerous alternative classifications. Degrees of similarity and dissimilarity, in homologous and to some extent in parallel characters but no others, then come in as secondary criteria for selection among the alternatives.

Criteria derived from relative divergence, that is, the sum of dissimilarities in all the characteristics studied, apply rather to the ranking of taxa already recognized than to their recognition in the first place. They may apply either within or between taxa or both, depending on the circumstances. The suggested criteria are as follows:

In a group of related taxa, members of the same category C_j, it is desirable that sequential divergences between least divergent pairs of T_js should be approximately equal.

In a group of related taxa of the same rank, it is desirable that the divergence within each of the T_js should as nearly as practicable approximate the average for the whole group.

In a hierarchic arrangement of related or inclusive taxa of different ranks, it is desirable that the divergences both within and between members of C_j should be less than in members of C_{j+1} and more than in members of C_{j-1}.

These are recommendations applicable in appropriate circumstances rather than general rules, because their application is sometimes impossible and can always be overridden by other criteria. The first recommendation obviously does not apply to a T_j in a monotypic T_{j+1}, for example aardvarks, a single family Orycteropodidae in the order Tubulidentata, where there is no familial (T_j) divergence. It is, however, the second rule that leads to ranking of the order. The divergence of aardvarks from other mammals approximates the average divergence of all usually recognized mammalian orders. It is that situation that justifies, or in some instances requires, the recognition of monotypic taxa. The absence of divergence at the family level also makes the third recommendation inapplicable as between order and family.

A possible fourth recommendation would be that in T_js included in the same T_{j+1} the divergence within each T_j should be less than that between T_{j+1}s. There are, however, many instances where *total* divergence within a well-defined genus, say, such as the exuberant *Rattus*, is greater than the average divergence between that genus and a closely related genus, such as *Thallomys* in the case of *Rattus*. Such instances are so numerous and in principle so likely to arise that the recommendation probably is not practical or desirable.

The accepted recommendations are simply more precise expressions of the general feeling among classifiers that taxa of the same rank should be of about the same "size," using "size" in the sense of morphological or other characterological scope, while taxa of higher rank should be of larger "size," in the same sense. On that basis, the import of the recommendations seems clear enough and probably requires no further discussion. There is, however, another meaning of "size" also pertinent in classification: the number of T_{j-1}s included in a T_j, that is, the amount of diversity in a taxon or the extent to which it is polytypic.

Here any recommendation must apply not only after the more impelling evolutionary criteria, such as that of relative monophyly, but also after the recommendations regarding divergence. There are enormous differences in diversity even in taxa with about the same internal and external degrees of divergence. *Eropeplus* includes only one named taxon in current classifications; the closely related *Rattus* includes more than 550, and there are genera even more diverse. In such examples the same generic rank given to taxa so different in diversity is based mainly on degree of divergence. Taxa that are extreme in diversity, either extremely low (especially monotypic) or extremely high (exceptionally polytypic) should nevertheless be carefully scrutinized in this respect. A monotypic genus, for example, in a given classification, such as *Asinus* in some classifications of recent Equidae, may represent an unusually and perhaps unjustifiably low criterion of generic divergence. If so, another classifier may enlarge the criterion of divergence, bringing two or more related proposed genera into one genus (*Equus* in the case of *Asinus*), which thus incidentally reaches a more usual level of polytypy. That may, however, be out of the question. No other recent genus could reasonably be placed near enough *Orycteropus* to make its family polytypic.

On the other hand, when extreme polytypy is present in a classification, consideration may be given to lowering the criterion of divergence and splitting the taxon into two or more of the same rank, each consequently less diverse. That is largely a matter of the possible convenience of having less unwieldy taxa to deal with. For example, much the most diverse group of mammals is that comprising, in the vernacular, the rats and mice and their closer relatives. Classically they were all referred to one family, Muridae, which was extremely unwieldy. It has now become fairly general practice to divide them into two families, Muridae and Cricetidae, both still very diverse, even though the degree of divergence is admittedly less than the average for mammalian or even for rodent families.[6]

In cases of extreme polytypy, first consideration should be given to making the taxon less unwieldy by use of intermediate lower taxa or subgroups. That is not an entirely convenient solution for the Muridae and Cricetidae, because even when separated they require the use of subfamilies to keep them manageable. There are, however, many highly diverse genera, for example, that are nevertheless fairly uniform (little divergent internally) and that can much better be subdivided into subgenera than split into different genera—*Rattus* is a good example.[7]

There is a natural and, within limits, justifiable tendency to use lower criteria of divergence in highly diverse groups and higher criteria in less diverse groups. Carnivora, for example, are much less diverse than rodents, and in most classifications the average divergence between carnivore families is notably greater than between rodent families. There are, however, other considerations that enter in. On present evidence, rodent families almost have to be less divergent if they are also to meet reasonable standards of monophyly.

SPLITTING AND LUMPING

Since the time of Linnaeus there has been an enormous increase in the number of recognized taxa at all stated levels. That has been in part a direct effect of new discoveries. Linnaeus had records of only a small fraction of recent and virtually none of extinct animals.

[6] Many genera formerly included in the Muridae are now also placed in still other families, but largely through recognition of other relationships or from changing concepts of monophyly.

[7] For various reasons, mostly psychological, many classifiers have an aversion to subgenera. In highly polytypic groups they are nevertheless so convenient as to be almost a practical necessity. Fortunately the aversion does not seem to extend to other intermediate taxa such as subfamilies.

Increase due to discovery was kept manageable partly by inserting new categories into the hierarchy, now commonly twenty or more within the animal kingdom in comparison with only four in Linnaeus. There has also been a secondary effect: as the known diversity increased the customary criteria of divergence at any given rank were narrowed, following the tendency noted in the preceding paragraph. That of course added to the number of taxa at what was *nominally* the same level. It may, however, be more clearly understood as a shifting of categorical names from one level to another. Thus it has often been noted that practically all of Linnaeus's animal genera are now families, at least, and some of them orders or taxa of even higher categories. The number of species increased so inordinately that many a Linnaean genus, if retained as such, would have hundreds, thousands, or tens of thousands of species. Merely intercalating superspecies and subgenera, even if that had been done consistently, which it was not, could not take care of the situation. So in effect what was done was simply to move the name of Linnaeus's categories downward in the scale, placing the name "family" where he had "genus," "order" where he had "family," and so on.[8] That gave the needed additional steps between, say, a Linnaean species and a Linnaean genus without changing the time-hallowed categorical names but just changing what they mean, their position in the scale. Or, to put the matter in a different way, the average degree of divergence that Linnaeus associated with the word "genus" came to be associated with the word "family" or "order."

Such adjustments most strongly affected lower levels of the hierarchy, down near the level of species where most of the expansion was occurring. Among recent animals, most of the higher taxa were already known to Linnaeus and adjustment at those levels came rather rapidly. Only 6 of the 118 families of recent mammals in my classification of 1945 were unknown to de Blainville in 1834, although of course his ranks were not necessarily the same. (See Simpson, 1954.) In the last century the addition of genuinely new higher categories has been due almost entirely to paleontology and has as a rule demanded intercalations into, rather than changes in, the ranks of cur-

[8] Of course it was not done in so orderly a way, as classification was changing also in other respects.

rent classifications. (The slow effects of increasingly evolutionary taxonomy have finally become more radical, but that is aside from the matter of relative ranks here under discussion.)

In addition to those adjustments of categorical levels to accommodate discoveries of hitherto unknown animals, there have been marked fluctuations in numbers of taxa caused not by discovery but by varying criteria among different classifiers. At the species level, typological approaches and rule-of-thumb criteria led some classifiers to define and name as species innumerable polymorphs and other distinguishable variants within populations of specific rank, at most, by phylogenetic standards. That tendency persisted late and is still not entirely corrected even among those whose approach is intended, at least, to be evolutionary.[9] Application of phylogenetic criteria has markedly decreased the number of species so created. (Here it is appropriate to say that the "species" were created by the classifiers, not by nature.) Modern revisions of long-known groups perforce list dozens or hundreds of specific synonyms, based on what is now recognized as intraspecific variation.

Despite many cases doubtful in principle or inadequately supported by needed data, phylogenetic criteria (next chapter) tend to produce fixed numbers of species comparatively independent of the personalities of the classifiers. Above the specific level, however, classifications with greatly different numbers of taxa of each rank may all have the necessary minimal consistency with the same inferred phylogeny. (Figure 10.) Here the numbers of taxa accepted are not so much a matter of generally accepted scientific principles of taxonomy as of personal taste in the artistic canon of classification. Some classifiers, the extreme splitters, prefer the narrowest practicable standards of divergence and diversity and consequently tend to recognize maximum numbers of taxa at each level. Others, the extreme lumpers, prefer the widest practicable standards and recognize minimum numbers of taxa. Splitters have proposed at least 28 genera of recent cats

[9] Paleontologists have been and to sharply decreasing extent still are among the worst offenders in this respect. I was taught in college that any fossil 15 per cent larger or smaller than the type of a named species could safely be placed in a new species. But some neontologists were no better. An eminent mammalogist in the same period told me that the best possible specific character is 100 miles (i.e., that specimens from two localities 100 miles apart are always to be placed in different species).

(Felidae), which may be even more than one genus per species. Lumpers have placed them all in one genus, or at most two genera (*Felis* and *Acinonyx*, the latter for the markedly aberrant cheetah).

Splitting and lumping are to some extent correlated with the purposes of the taxonomist. One who is concerned primarily with sorting and identifying specialized collections into species naturally has

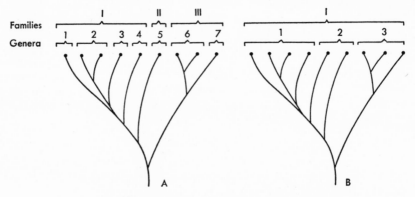

FIGURE 10. DIAGRAM OF SPLITTING AND LUMPING

A phylogeny (or dendrogram) of nine recent species (black dots) is postulated as accepted by both classifiers.

A. Split classification with seven genera and three families.

B. Lumped classification with three genera and one family. The genera of a lumper may be the same taxa (only at a different level) as the families of a splitter, but this diagram shows that they may be different taxa even when the same in number. Both classifications are fully consistent with the postulated phylogeny. Several other consistent arrangements are also possible.

his attention focused on dissimilarities and on diagnosis—telling taxa apart.[10] His approach is analytical, and he is likely to be a splitter, at least in his own special group, which he knows so well that a multiplicity of supraspecific taxa seems more of a convenience than a nuisance. (Many a taxonomist who is a splitter in his specialty wishes that others would be lumpers in theirs!) On the other hand, a taxonomist concerned with building up a hierarchic system for major

[10] "Diagnosis" is ultimately from Greek διαγιγνώσκω, "distinguish between [two] things," and "definition" from Latin *definio*, "enclose within limits." The concepts are importantly different, but in fact taxonomists seldom use the terms consistently and commonly assume that they are synonymous.

groups of animals must have a more synthetic approach and must focus to considerable extent on the similarities that permit recognition of collective taxa and on their definition—delimiting their collective boundaries rather than only distinguishing them from other taxa. He will tend to be a lumper. In addition to these more readily understandable differences in purpose there are, however, more obscure personality differences between splitters and lumpers (Roe, 1953). Some people just are more inclined to take things apart and others to put them together.

As usual, there are arguments in favor of each extreme and there is room for more moderate artistry between the two. The following indirect dialogue between J. A. Allen, an extreme splitter, and me, frankly a lumper but not, I trust, an extreme one, brings out main points of the polemic.

Allen (1919): These groups are given, tentatively and for convenience, the nomenclatural rank of genera. Their evaluation by future investigators will be subject to general equation, from the viewpoint of whether all cats should be placed in *Felis*, because they are neither bears nor wolves, or that a few leading types may be segregated, to show that the domestic pussy is neither a lion, nor quite a leopard, nor a lynx, or whether it is better to indicate that even among the smaller cats there is some diversity of structure and relationship. To illustrate, the latest faunal list of North American Felidae includes forty-one forms, of which sixteen are placed in the genus *Lynx* and twenty-five in the genus *Felis*. Of course the merest tyro in natural history will see at a glance that a part are lynxes and the rest are cats, but only a specialist in cat nomenclature will be able to recognize which are big cats (as pumas and jaguars) or small cats, or which are plain-colored and which are elaborately patterned with spots and bands, or what are their natural relationships; except in the case of subspecies where the trinomial is the key. A further use of generic divisions would not obscure the fact that they are all cats but would indicate that at least all cats are not alike, and perhaps inspire interest as to how they differ.

Simpson (1945): This argument is almost as disingenuous as it is ingenious. In the first place, its whole point of view is that the purpose of classification is to separate things, barely hinting at the vital fact that bringing things together properly is the more useful and important function of taxonomy. In the second place, Allen seems to deplore the existence of any taxonomic grades between subspecies and genera. In fact a careful examination of his classification shows that he does not use the specific grade in any useful way; in his system the genus does the work and takes the place of the species and the next smaller working unit is the "form," subspecies, or local race. As for subgenera, he ignores them altogether. Thus while insisting on further generic splitting in order to show grades of differences (and, in theory but not in his practice, of resemblances), he throws out the very mechanism by which this can best be done. It is virtually impossible to show such grades if generic, subgeneric, and specific ranks are all leveled out as genera.

Perhaps it is not necessary to emphasize the especially disingenuous touches in this argument. Allen surely well knew that the fact that cats are not bears or wolves is shown (by everyone) by their being placed in a different superfamily, not by their generic allocations. It is also impossible to believe that Allen really thought that it would help "the tyro in natural history" to make him learn seven generic names in place of one, or that, for instance, there is anything in "*Margay glaucula glaucula*" that will convey to the tyro the idea of a small, gray, spotted Mexican cat more clearly than would *Felis tigrina glaucula*. Indeed a tyro would never guess, and not all specialists would know offhand, even that "*Margay*" is a cat, and the name used by Allen shows nothing as to its relationships, whereas the other name reveals at once to the specialist and also in less degree to the tyro a whole sequence of affinities in different degrees.

There will continue to be splitters and lumpers, and the personality factor can no more be eliminated from classification than from any other art. There is, however, a reasonable canon that still gives leeway for that factor. It is in the nature of things that some monotypic taxa are necessary in consistent, well-balanced classification. If, however, in a classification a considerable number of T_js are monotypic, consideration should be given to the probability either (a) that the T_{j+1}s or higher taxa are too split, or (b) that the T_{j-1}s or lower taxa are too lumped. More generally, a classification well balanced as to splitting and lumping will have, as an average throughout the classification, a considerably higher number of T_{j-1}s than of T_js for all levels of j. In this respect, too, attention must be given to the desirability of stability (avoid unnecessary changes) and of convenience (think of the users of classification who are not specialists in the given group).

RELATIVE ANTIQUITY

In accepted evolutionary classifications every T_{j+1} is at least as old as any of its included T_js and, as a rule with possible but certainly rare exceptions, each T_{j+1} is older than the average age of its included

T_js. Those are necessary and, it would seem, obvious consequences of the nature of phylogeny and the arrangement of evolutionary taxa in a hierarchic system. The statement is as important for what it does not say as for what it does say. Among the things that it does *not* say are these:

1. T_{j+1}s are at least as old as and usually older than T_js. (The relationship holds only for a *given* T_{j+1} and *its included* T_js.)

2. T_{j+1}s may or do evolve before their included T_js. (The relationship is valid only for the *average*, not for all *individual*, ages of the T_js.)

3. T_js tend to be, or are, of the same age. (Nothing in the statement is logically related to that conclusion.)

Those and similar misapprehensions have sometimes confused phylogenetic classifiers and have also been advanced occasionally as arguments against the practicability of phylogenetic classification. The first does represent a tendency or a usual average for any given group, but it could not be strictly or generally true unless rates of evolution were uniform, which is contrary to known fact. Bigelow (1958), one of the rare thoughtful opponents of evolutionary classification, has argued that phylogenetic classification demands uniformity of evolutionary rates, but that is a straw man as Bader (1958) has conclusively demonstrated in a reply to Bigelow. As has been discussed previously in this chapter, evolutionary classification can and does take account of degrees of divergence and is not on that account any less consistent with phylogeny. A more rapidly evolving group will diverge more than a more slowly evolving one and may for that reason be given higher categorical rank. The rapidly evolving group of elephants (and mammoths, which are merely extinct elephants) is so divergent as to warrant family rank, but it is not as old as the likewise proboscidean, slowly evolving genus *Mammut* (the so-called American mastodons).

Nevertheless, within any one group (of almost any scope or rank) the T_{j+1}s do tend *on an average* to be older than the T_js. All concrete estimates of actual evolutionary rates that have been made (for example, Simpson, 1953) show a very strongly peaked mode, with modal (in that sense, average) rates much more common than notably faster or slower rates. That explains the tendency for T_{j+1}s to be older than T_js. From one group to another, however, the modal rates may be very different, so the tendency cannot be generalized in that way.

That higher taxa evolve before included lower taxa has been imputed to phylogenetic classification as a false premise, but is actually accepted by few if any evolutionary classifiers as the terms are used in this book. This would, in a sense, be possible if higher taxa originated in single mutations, as maintained by Schindewolf (for example, 1950), who is of course an evolutionist but whose taxonomy is essentially typological. Even in a system of taxonomy like his, the statement obviously cannot have been literally true when the higher taxon originated. All organisms belong to species, not only by the accepted definition of species but also in the nature of things by any definition of the category species. However a higher taxon may have arisen, its first members belong to one or more species, which therefore are coeval with the first genus, which is coeval with the first family, and so on for origins of any higher taxon. The original lower taxa of a higher taxon tend sooner or later to be replaced by other lower taxa, so that eventually no lower taxa in existence *at a given time* may be as old as a higher taxon in which they are included. For contemporaneous animals, such as the recent fauna, it may then be and in fact usually is true that no *known* lower taxon is as old as the higher taxa including it—and of course only known taxa actually figure in a classification. That does not alter the fact that the higher taxon cannot be older than included groups that did once exist and that would be classified as lower taxa if known.

Another clarifying way to look at this matter is to look at taxa as symbolized by parts of a phylogenetic tree. Unless higher taxa commonly evolve in one step, which almost all evolutionists think extremely improbable, they will then represent larger sections of the tree than their included lower taxa, and in general the average sizes of taxa on the diagram will be proportionate to their ranks. A higher taxon then simply does not exist as such until the correspondingly larger section of the tree has evolved by prolification of lower taxa. (See Figure 11.) In terms of the evolving pattern then, the higher taxa appear only in retrospect *after* the successive lower taxa composing them, and not the reverse.

That taxa of the same rank, whether within any one higher taxon or in general, should be of the same age is absolutely impossible in any reasonable system of classification. It would seem conceivable that such a criterion could be applied to *contemporaneous* animals only, and attempts along those lines have actually been made. They

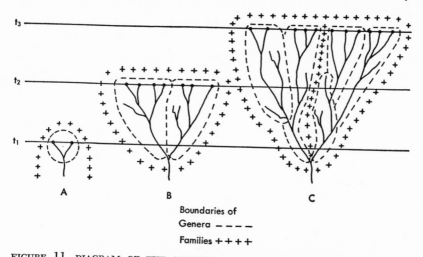

FIGURE 11. DIAGRAM OF THE GROWTH OF A PHYLOGENETIC TREE AND ITS RELATIONSHIP TO THE ORIGIN OF HIGHER TAXA

The horizontal lines, t_1, t_2, and t_3, symbolize three successive time levels. The heavy dots are species existing at one of the three times. (This symbolization is adopted to clarify the situation and its application to contemporaneous species; in reality each species also has a vertical time extension and is a segment of the phylogenetic line, not a point on it.) Broken lines enclose genera, and crosses enclose families, each as of one of the three successive times.

A. Two species at t_1, comprising (at most) a genus of an older family.

B. Seven species and two genera at t_2, comprising a family that did not exist at t_1, although the genus ancestral to it, and now included in it as parts of two different genera, existed then. The two ancestral species are now placed in the two different genera that became such *after* those species lived. The family segment is larger than the generic segments.

C. Eleven species, four genera, and two families exist at time t_3 as a continuation of the processes that evolved B from A.

turn out to be completely impractical, however. Within the vertebrates the classes (as now ranked) Osteichthyes, Amphibia, Reptilia, and Mammalia are successively and markedly younger in that sequence. By a criterion of equal antiquity they should therefore have successively lower ranks, and supposing that Osteichthyes were retained as a class, Mammalia would be a superorder, at most, or a genus if numerous intercalary ranks were not used. If the whole animal kingdom were so arranged, we would find ourselves with such taxa as an "order Latimeriae" (for what is now the fish genus *Latimeria*) and a "genus *Primas*" (for what is now the order Primates). To obtain a workable arrangement we would also need a greatly increased number of hierarchic ranks, probably two or three hundred if not more. All of which could be done, but at an impossible cost in clarity and convenience and without need or justification in the principles of evolutionary classification.

In any event, such a criterion could be consistently applied only to contemporaneous animals. If temporal sequences were brought in, the same kind of biological unit that we call a species in the Recent would have to be called a phylum, at least, in the early Cambrian, and the whole hierarchic system would break down.

It is thus evident that a general criterion of relative antiquity is not and practically cannot be applied to the ranking of evolutionary taxa. It is still true that *within* a group (higher taxon) of *contemporaneous* animals a criterion of relative antiquity is justified in principle and useful in practice. The case is a special one, but it is important because most classifications do relate to defined groups of contemporaneous animals chiefly if not exclusively. The appropriate criterion is simply a reflection of the relationships stated in the first sentence of this section:

Taxa of contemporaneous animals should be ranked in such a way that the common ancestry of every T_{j+1} occurred as early as that of any of the included T_js and earlier than the average for the T_js.

There are several other lines of evidence on relative antiquity (for example, crossing specializations, previously mentioned), but most important is that of over-all similarity in homologous or, with greater caution, parallel characteristics. As a rule, two (or more) taxa more similar in such characteristics had a more recent common ancestry (have closer propinquity of descent) than two (or more) other taxa, *one of which includes the preceding two (or more)*, that are less similar. The italicized condition is based on the application of the criterion *within,* not *between,* groups. While omitting that essential qualification, Bigelow (1958) has vigorously objected to this line of evidence on grounds that no correlation of similarity and nearness of common ancestry is expectable in principle.[11] Bader (1958) has, however, conclusively answered the objection, largely on the basis of modal rate distributions as previously mentioned. In fact, recognition that the criterion applies to the ages of T_{j+1}s relative to their *included* T_js, and not among T_js, is itself enough to rob the objection of all force.

A second objection, also made by Bigelow and others, is that ages of ancestors cannot be surely determined from contemporaneous animals, alone. It is not, however, required that the absolute ages be established. Such inferences as are symbolized in dendrograms do not involve direct representation or reconstruction of the ancestry, but if they are based on adequate data interpreted on sound evolutionary principles they indicate the *relative* ages of groups with a sufficiently high degree of probability. (It is often helpful but is not necessary for this purpose that the inferences actually be drawn as dendrograms.)

RANKS OF CHARACTERS

The art of classification is concerned first with the delimitation and then with the ranking of taxa. It would be a delightful simplification if the characters of organisms had an inherent association with the various categorical ranks. That subject was treated sufficiently in previous chapters and need not here be discussed at any length, but it is mentioned again by way of emphasis and summary.

Some kinds of characters do have a loose association with taxonomic ranks, and that fact is of some help in sorting out taxa. It is not, however, in itself an acceptable criterion for ranking. It is populations of organisms that are classified, not their characters. A generic character is one that characterizes a genus, and it is not determinable a priori. Identical characters may in one group characterize a species (or even be a mere variation in a species; see Simpson, 1937), in another a genus, and in another a family.

[11] He also brings in the problem of convergence, here already sufficiently discussed.

It is unfortunate that organisms are not naturally grouped by simple relationships, simply arrived at, but we must take nature as we find it.

One final comment: In Chapter 2 much was made of the necessity for assigning priorities among the various relationships of animals and among the characteristics that reflect those relationships. The phylogenetic system, with its degrees of propinquity of descent from common ancestry, provides the basis for recognizing priorities that are literally prior, in the temporal sense, and that are certainly natural in the fullest sense of the word. (On this and some related topics, see the thought-provoking paper by Pirie, 1952.)

5

The Species and Lower Categories

Definition of the category species has long been one of the major problems of taxonomy, in part because of the incongruity of lingering typological concepts with the nature of evolving species and in part because evolutionary species may have diverse properties and are only gradually differentiated in phylogeny. The modern biological or, more strictly, genetical concept of a species among biparental, contemporaneous organisms is a group of interbreeding organisms genetically isolated from other such groups. That is a special case of a more extensive evolutionary concept of the species as a lineage with separate and unitary evolutionary role. The genetical species closely approximates a temporal cross section of such a lineage in biparental populations. With some practical qualifications the evolutionary definition also applies to uniparental populations and to those changing through time, and thus relates all to one general concept. Other concepts of species must be considered, but none are so suitable for strictly taxonomic use.

Species in time, as segments of phylogeny, raise special problems of subdivision of unbroken successions of species and treatment of branching lineages. The required procedures must be arbitrary and give scope to variations in the art of classification, but some practical criteria can be found.

For evolutionary study and descriptive purposes, many kinds of variation within species come into consideration, but the only regular infraspecific taxon is the subspecies. Subspecies are widely but not universally useful, and problems of their definition yield to the same sorts of criteria as for dividing other phylogenetic continua.

Proposed usage of superspecies illustrates some relationships and differences of general interest between evolutionary concepts and the taxa of classification. Finally, the relationship of samples to populations and their implications for classification again require comment in the context of this chapter.

THE GENETICAL SPECIES

To the ancients and to the scholastics species were neither immutable nor sharply delimited. Along with a belief in the spontaneous generation of many highly organized animals (for example, bees, rats), went beliefs in erratic mutations, radical degeneration,[1] and monstrous hybridization between almost any animals no matter how discrepant (for example, camels and sparrows). Such order as was seen among organisms seemed to occur not because of but in spite of the changeability of species—a concept still retained in Lamarck's abortive evolutionism. Moreover, the most conspicuous pre-evolutionary theory of order, that of the *scala naturae,* usually assumed that all species formed a continuum and were inherently arbitrary and nondelimitable. Dogmatic belief in fixed and sharply distinct species was a relatively late theological development and had not long been accepted by pious naturalists when Linnaeus wrote his famous aphorism that there are as many species as were created in the beginning. That would have shocked Aquinas, for example, who stated emphatically and explicitly that new species are constantly being produced (but not by evolution!). (The remarkable pre-evolutionary change in the species concept has been well discussed by Zirkle, 1959.)

For a relatively brief halcyon period taxonomists could and did believe that an exact number of clear-cut species exists in nature. The rest of the hierarchy presented many problems, but recognition and enumeration of species seemed in principle to be a simple task. In that atmosphere, proof of evolution amounted to demonstration that one species can give rise to another. That is why Darwin's book, which

[1] Up until almost modern times "degeneration" meant production of descendants outside the parental or ancestral norm and did not necessarily indicate deterioration.

treats all levels of evolution, and not species alone, was called *The Origin of Species*. The habit of mind has persisted, both among anti-evolutionary sectarians and among evolutionary biologists, and speciation continues to be so crucial a point as frequently to be considered virtually synonymous with evolution.

Apart from that point, post-Darwinian taxonomists observed that there are units in nature that have a special evolutionary status not fully shared with taxa either above or below them in the hierarchy. The units are frequently so distinct that they can be recognized without knowledge of their real nature. Many of them so recognized before Darwin had been called species, and it was inevitable that the term should be transferred to the evolutionary units. In most cases among earlier evolutionists it was not even realized that the term was being transferred to a new concept, and there was an ambiguous tendency to *think* of the species as an evolutionary unit but to try to *define* it typologically. Some taxonomists still do not understand or do not accept the distinction, and the continuing ambiguity is one of the reasons for the "species problem"—the problem, which has probably caused more ink to flow than any other one point in taxonomy, is to produce a generally acceptable and workable definition of the category called species.

There are three other main reasons why that is a problem. All arise from the nature of evolution itself, and they make it impossible ever to produce the kind of solution that some taxonomists have naïvely demanded. There can be no one definition of the species applicable to all organisms and unambiguous in its application to every individual case. The three reasons are:

1. Members of a category around the level of species are not invariably separable into groups by any absolute criterion.

2. Groups recognizable as species may yet differ in population structures and mode of origin so that their category cannot be adequately defined by the same evolutionary criteria for all cases.

3. It is universal and seemingly inevitable practice to use the same categories for temporal subdivisions of phylogenies as for groups of contemporaneous animals, but the resulting taxa differ in kind so that application of a single categorical definition to both is difficult, at least.

At this point I shall assume, for reasons already amply given, that acceptable modern definitions of the species category should have a meaningful relationship to evolution. Much of the rest of this chapter is devoted to some of the problems that arise because species are in fact evolutionary units. I shall not attempt to cite much of the enormous literature, but draw special attention to two fairly recent symposia (Sylvester-Bradley, 1956; Mayr, 1957), to serial discussion in the journal *Evolution* (especially the papers cited and in part summarized in Simpson, 1951), and to the useful book by Cain (1954).

The definitions of the species now most widely accepted are variants or equivalents of that by Mayr (1940, also many of Mayr's later papers and books):

Species are groups of actually or potentially interbreeding natural populations, which are reproductively isolated from other such groups.

Although often called the biological species concept or definition,[2] that is strictly a genetical concept of species among contemporaneous, biparental organisms. Its meaning is quite clear, and an experienced taxonomist can apply it with little difficulty to probably at least nine tenths, perhaps even more, of animals to be classified. In most groups of recent animals it is both applicable and sufficient, and it is accepted and recommended on that basis. It has, nevertheless limitations and difficulties, which are of course well known to Mayr and its other proponents. In the first place, it does not apply, even in principle, to temporally sequential species or to species of uniparental organisms. Those special (but very common) cases, for which the genetical definition is simply irrelevant, will be discussed later. The main difficulties, some real and some only apparent, will now be considered.

The definition depends on a criterion, interbreeding, that usually is not and sometimes cannot be observed. That is only a pseudoproblem, as the previous discussion of the relationship between definition and evidence in taxonomy should make clear. The definition does define, and does so in a biologically, evolutionarily meaningful way. The evidence that the definition is met in a given case with a sufficient degree of probability is a different matter. The evidence is usually morphological, but to conclude that one therefore is using or should use a mor-

---

[2] A mildly misleading designation, because the many additional or alternative definitions are equally biological. In fact I do not see how any *relevant* definition for a category of groups of organisms can be anything but biological.

phological concept of the category (not taxon) species is either a confusion in thought or an unjustified relapse into typology. The evidence is to be judged in the light of known consequences of the genetical situation stated in the definition. Not the only but the most important single practical criterion is as follows:

If the ranges of population variation (including polymorphy) inferred from two or more samples overlap for all observable characters, there is high probability that the corresponding populations were or had recently been interbreeding when the specimens were alive and that they therefore belong to the same genetical species. If there is significant absence of overlap in inferred ranges, the populations probably had not recently been directly interbreeding and may belong to different genetical species. If, however, the samples are from more or less widely separate geographic localities, one must consider and as far as practicable test the possibility that the populations in question were both interbreeding with intervening populations by which they were united into a single genetical species.

A second, real but readily superable difficulty is the vagueness and the difficulty of testing the concept of potentiality for interbreeding. If, as frequently happens, populations are disjunctive, it is improbable that much if any interbreeding is actually going on, and its extent will depend on frequency of movement of individuals (or in some aquatic animals of gametes) from one area to the other, which may not be ascertainable. If no appreciable genetic differences have developed among them, it would be inconvenient or almost ridiculous to insist that each disjunct population is a separate genetical species even if no interbreeding is occurring. In such cases the same criteria of overlap may reasonably, although more or less arbitrarily, be applied. Presence of overlap shows that interbreeding had recently occurred and makes a *prima facie* case for its continuing potentiality. Absence of overlap suggests definite genetical divergence and in the absence of intervening populations establishes a reasonable doubt (although it cannot disprove) that interbreeding should still be considered potential for purposes of definition.

A third difficulty, the only one that is really serious when the definition is fully relevant, relates to the degree of isolation. A few taxonomists have insisted on absolute permanent isolation: the impossibility of production of fertile hybrids. Most taxonomists hold that condition to be sufficient but not necessary: populations that cannot produce fertile hybrids (either directly or through intermediate populations) are *ipso facto* specifically distinct, but populations that can produce fertile hybrids are not *ipso facto* conspecific. The more stringent criterion is undesirable, because populations that are in all other respects exactly like unquestionable species do occasionally produce fertile hybrids. It is also impractical, because the possibility of producing fertile hybrids can rarely be judged adequately on available evidence except for species so obviously distinct that the question need not be raised at all.

Species do evolve, and almost always do so gradually. Among evolutionary species there cannot possibly be a general dichotomy between free interbreeding and no interbreeding. Every intermediate stage occurs, and there is no practically definable point in time when two infraspecific populations suddenly become separate species. Fortunately for the neontologists, the majority of living populations have either definitely passed that hypothetical point or are not yet close to it. Nevertheless speciation is actively occurring today, and many populations are in the intermediate stages of some, but reduced, interbreeding. Again, if there are distinct gaps between ranges of characters, it is sufficiently probable that isolation is at least complete enough to warrant specific separation. There remain numerous doubtful cases where decision depends on the personal judgment of each practitioner of the art of classification. To insist on an absolute objective criterion would be to deny the facts of life, especially the inescapable fact of evolution.

THE EVOLUTIONARY SPECIES

It is the fact of evolution that has made genetical species separate and that keeps them from always being sharply, clearly separate. It is also evident that the genetical definition of species has evolutionary significance. Still it is striking that the definition does not actually involve any evolutionary criterion or say anything about evolution. It would apply equally well, or in fact a great deal better, to species that did not evolve. Mayr (1957) has pointed out that pre-evolutionary taxonomists advanced almost identical definitions, for example, Voigt in 1817: "Whatever interbreeds fertilely and reproduces is called a species." [3]

Given the fact that the genetical definition of species is consistent with evolution, its lack of any direct and overt evolutionary element certainly does not invalidate it. Nevertheless it is desirable also to have a broader theoretical definition that relates the genetical species directly to the evolutionary processes that produce it. I have elsewhere (Simpson, 1951) proposed such a definition, which may be slightly modified as follows:

An evolutionary species is a lineage (an ancestral-descendant sequence of populations) evolving separately from others and with its own unitary evolutionary role and tendencies.

That definition not only is consistent with the genetical definition but also helps to clarify it and to remove some of its limitations. At any one point in time, or taken in temporal cross section, an evolutionary species of biparental organisms will almost invariably coincide with a genetical species. That is precisely why the genetical species has evolutionary significance, and in a broader sense it is why genetical species do exist. It also tends to remove some of the conceptual difficulties remaining in the purely genetical concept. It is (to me, at least) clearer to see why disjunct populations should be placed in one species if they retain the same evolutionary role than if they have the rather vague potentiality to interbreed. It is also clear that two species may interbreed to some extent without losing their distinction in evolutionary roles and that this is the really important point for evolutionary taxonomy. The amount of interbreeding allowable by definition is then precisely as much as does not cause their roles to merge. The taxonomic value of the genetic criteria of interbreeding and isolation lies not in those characteristics in themselves but in their evidence as to whether populations are or are not capable of sustaining separate and unitary roles over considerable periods of time. Interbreeding helps to keep a role unified; isolation makes possible separation of roles.

The evolutionary definition given above omits the criterion of interbreeding. Interbreeding promotes the unity of role involved and is also evidence that the criterion of unity is met. It is not, however, the only way in which unity is maintained or the only evidence that it

[3] "Mann nennt Spezies . . . was sich fruchtbar mit einander gattet, fortpflanzt." Citation and quotation from Mayr.

exists. The evolutionary definition is thus broader than the genetical, while including it as the most important special case. Uniparental populations also have separate and unitary roles, and the evolutionary definitions can also be applied to them, as discussed later in this chapter.

The evolutionary species implicitly brings in the element of time. Species do in fact have a long time dimension, and a concept that omits this consideration is incomplete if not quite inappropriate. The evolutionary concept is thus more readily related to paleontological sequences, a point also discussed farther along in this chapter.

The one important difficulty in the evolutionary concept of the species is the definition and recognition of roles. That is rarely a serious problem to a field naturalist or ecologist, who can almost always see clearly that the various species he encounters do have recognizably different roles. Roles are definable by their equivalence to niches, using "niche" for the whole way of life or relationship to the environment of a population of animals and not for its microgeographic situation.

Here again the question of definition and evidence comes up. The role cannot be directly observed in a series of dead specimens, recent or fossil, in a museum. Valid and sufficient evidence of separation and unity in roles can, however, be obtained from observation on such specimens. Morphological resemblances and differences (as reflected in *populations,* not individuals) are related to roles if they are adaptive in nature. The assumption that over-all resemblance and difference is, on balance or as an average, adaptive is adequately justified by general evolutionary theory. It is helpful but it is not necessary for purely taxonomic purposes that specification of the nature of the adaptation, or of the role, be possible. The definition requires only that the roles be separate but each unified, and that is, as a safe enough rule, shown by somatic differences and resemblances between populations. That may at first sight look like a circular procedure, but it is not. The normal correlation between role and morphology has been amply established for populations in which the roles were independently determined on quite different grounds, notably those of environment and behavior. Another check and further aid in applying criteria as to roles is, moreover, provided by adequate field data, which even for fossils yield at least some information on environment.

Infraspecific groups may differ somewhat in roles. As a rule, however, these differences, like morphological differences, grade through adjacent populations, and the criterion of distinct separateness of roles would not make such groups different species. Again there are and in the nature of things must be doubtful cases, for nascent species do not suddenly acquire separate roles at a determinable instant.

The definition neither states nor implies that the unitary role of any one species is necessarily unchanging. Evolution (apart from quite exceptional saltations) could not occur if roles did not change *within* species. The concept involves a species having a unified role at any one time (not necessarily the same role at all times) and that its role always be separate in some way (not always in the same way) throughout its duration.

SOME OTHER KINDS OF SPECIES

Several other definitions of species, not merely variants of the preceding two, are in current use, and special names have been proposed for a number of them. I shall not attempt any exhaustive account of them, but shall mention a few either because they involve some interesting taxonomic concepts or because they are frequently mentioned in taxonomic literature. Cain (1954) distinguishes and names the following kinds of species:

Taxonomic species: a general expression for any taxon that has been called a species and given a specific name available under the International Rules of Nomenclature.

Morphospecies: established by morphological similarity regardless of other considerations. These are discussed further in this section.

Palaeospecies: [4] temporally successive species in a single lineage. These are discussed in a later section.

Biospecies: [5] the concept I have called genetical species and have already discussed.

Agamospecies: [6] species of uniparental organisms. They are also discussed in a separate later section.

[4] The name is not ideal, because the great majority of paleontological species are not "palaeospecies" by this definition. It is, however, impossible to make etymology and definition precisely equivalent, and if a clear definition is given, the etymology has little importance unless it is drastically misleading.
[5] For "biological species," a term to which I have already mildly objected.
[6] Also somewhat misleading, because the point is not that they lack gametes but that they are uniparental.

The subject of "morphospecies," including what paleontologists often call "form species" (plus "form genera," etc.) and "paraspecies" (or "parataxa" in general), has become highly confused. The terms have been used in many different senses, often poorly defined or not defined at all. The implication and sometimes the definition of morphospecies and form species is that they "have been established solely on morphological evidence" (Cain, 1954). That is, in my opinion, quite unsatisfactory. The species concept applies to groups of whole organisms. If the definition literally means that *nothing* but morphology is taken into consideration, the classification would be one of characters and not of organisms. The concept might apply, although even there not without some qualification, to some typological or strictly empirical concepts of species, but not to any concept acceptable in evolutionary taxonomy. The most important point here is not so much the evidence that is used as how it is interpreted. Genetical and evolutionary species are usually based largely or entirely on morphological evidence. It seems to me to be confusing to bring them on that account under a rubric of "morphospecies," along with numerous other and conflicting concepts. It has certainly confused a few taxonomists who argue that *because* most species, as taxa, are based on morphological evidence the category species *cannot* have valid genetical or evolutionary definitions.

"Form species" or "morphological species" have frequently been erected and named by stratigraphic paleontologists by selecting arbitrary "types" among specimens suspected or positively known to belong to single biological populations or "biospecies." This return to pure typology has been ardently defended, even in recent years (for example, Eagar, 1956), chiefly on the grounds that it is helpful in stratigraphic correlation. Now, no one can object to stratigraphers' devising any system that they find useful within their own special procedures. Some practicing stratigraphers do in fact correlate by fossils that they do not classify, name, or even consider as biological objects, and there is no possible complaint against such practice. There are, however, legitimate and violent objections when they make admittedly nonbiological groups and then classify them as if they were biological and give them Linnaean names with which biological taxonomists are forced to deal, if only as synonyms, under the International Rules. In fact such stratigraphers are defeating their own purported aims, because it is demonstrable that well-analyzed evolutionary species permit closer and more reliable correlation than do "morphospecies" (for example, George, 1956).

Still another concept of form taxa, also most common among paleontologists,[7] is that of polyphyletic groups, united either by mistake or for convenience by parallel or convergent morphological characters. Form genera (or higher taxa) in this sense are more common than form species but the concept is essentially the same. The previous discussion of grades and clades is also pertinent here. I can see no reason for the deliberate use of form taxa when they are inferred not to be monophyletic by a reasonable criterion of monophyly. Recognition of the existence and nature of the parallelism or convergence, or of the grades observed, is a useful and enlightening result of taxonomic research. It does not, however, warrant insertion of such quite different kinds of groups into the usual system of classification, nor does it demand for its recognition and discussion the erection of a second system of classification and nomenclature. Certainly mistakes are made and "species" (and higher taxa) believed to be genetical or evolutionary when proposed are later found to be form taxa. But we hardly need or could apply a species definition or concept for groups that are some time in the future going to be recognized as not what we now think they are!

One of the problems most peculiar to paleontology is the necessity for somehow classifying parts or traces of organisms out of association with each other or with the whole organism. For example, conodonts, extremely useful horizon-markers, occurred in the whole animals as complex assemblages of markedly different small, hard plates, spines, combs, etc. Few assemblages have been found, and for the most part classification of conodonts has to be based on isolated parts of unknown relationship to each other. The inevitable result, as substantiated in a few cases by natural assemblages, is that different "species" and "genera" are routinely based on what were parts of one animal. (References and exemplifications in, for example, Scott and

[7] In specifying what I consider defects in the practices of some paleontologists, I do not imply that they characterize paleontology as a whole. They do not, and they are not confined to paleontologists.

Collinson, 1959.) Fossil footprints pose the same problem, because they almost never can be associated with lower taxa known from skeletons. (For example, Lull, 1953.)

One solution would of course be to refrain from giving Linnaean names to parts of animals but to designate them in some entirely different way. That has been tried in isolated instances, but it is not an acceptable general solution because there is no good place to draw the line. *All* of paleontology is based on parts, only, of animals and it would be a crippling blow to evolutionary biology if regular biological classification were not attempted, at least, in all cases. The usual current practice is to go ahead and classify unassociated parts, footprints, etc. as well as possible and to hope for the best: that discovery will eventually permit bringing them together into assembled animals. The same procedure has sometimes been used in neontology, for example when different species were based on different life stages of the same animals. Recently a compromise solution has been suggested in paleontology: that groups based on parts of conodonts and similar usually dissociated parts of fossil organisms be recognized as *parataxa* and form a system separate from the usual classification for nomenclatural purposes (Moore, 1957). The proposal was presented to the International Congress of Zoology in 1958, but was not given official recognition.

The fact that the genetical definition of the species does not mention morphology suggests the possibility that some good genetical species might not differ in anatomy. That is sometimes nearly true, at least, and pairs of anatomically closely similar genetical species are called *cryptic* or *sibling species*. (Extensive discussion and exemplification in Mayr, 1942, 1948.) Many such pairs have been found, although the known cases are an almost insignificant fraction of all animal species. Since most species have in fact been defined [8] only anatomically, there is some possibility that sibling species are really very common or practically universal in nature. If so, the genetical species, while retaining its theoretical interest, would be virtually valueless in the actual practice of classification. That conclusion has in fact been urged by Sonne-

---

[8] A species of organisms, a *taxon*, may be and commonly is defined anatomically (or morphologically) although the *category* of which it is a member is defined genetically. There is no real contradiction in this apparent anomaly, which is again related to the difference between definition and criteria.

born (1957), but on evidence of protists, which are in this respect not on a par with most metazoans, and with questionable force even as regards protists.[9]

For metazoans, at least, there is some reason to doubt whether this problem is quite as important as it seems at first sight. There are extremely few examples in which the species called siblings did not prove to be anatomically distinct when studied more carefully and with simultaneous consideration of all available anatomical characters. In fact the separation of so-called sibling species has in some cases been largely or solely on anatomical grounds. For example, the "sibling" stone and pine martens (*Martes foina* and *martes*) have been confused in attempts at definition by nondiagnostic characters but are unequivocally and immediately distinguishable when their multiple diagnostic characters are used. (Streuli, 1932; this is cited by Mayr, 1942, as one of the few examples of supposed sibling species in mammals.)

Lineages slowly diverging or speciating at first have minimal morphological distinction but in most cases, at least, the distinction increases constantly as time passes. It would seem that the stage of no or minimal distinction would usually *precede* the critical point of genetical isolation or would at most continue past that point only for a brief period. At any one time, comparatively few animals would be in just that equivocal state of established genetical isolation but little or no morphological distinction. It is probable that a number of species that differ in usual and distinct ways but to comparatively slight degrees (like the martens) are in that state. Application to them of a special sibling concept is of doubtful value.

There are, however, some analyzed examples in which genetical divergence is demonstrably great while anatomical divergence is hardly appreciable, for example, the famous case of *Drosophila pseudoobscura* and *persimilis* (Dobzhansky, 1951). This exemplifies a previous comment that the manifest relevance of genetic to somatic evolution is by no means an identity and may in particular instances show little correlation. If it is to be really meaningful, the term sibling species should, I think, be confined to such cases. By this more stringent definition

[9] It is a great but not completely unfair oversimplification of the argument to say that Sonneborn considers the recognition of sibling genetical species impractical because he has recognized many of them. It is further probable that sibling species are much more common in protists than in animals.

known sibling species are exceedingly rare. It seems probable that true sibling species are in fact rare, but that is to some extent an article of faith because so few adequate genetical analyses of natural species have been made. It is impossible to make such analyses routinely in general taxonomic procedure, and if it should happen that true sibling species are common, the genetical species concept will have greatly reduced value for classification.

Different genetical species that lack any determinable anatomical or ecological distinction are single species under the evolutionary definition: they do not have definably separate evolutionary roles. The collective evolutionary species so defined is analogous to uniparental species, in which unity of role is maintained without interbreeding. Applicability to such cases is an advantage of the evolutionary definition, which retains its usefulness in classification regardless of the existence of sibling species.

Following a different line of thought, Rensch (1929) distinguished between *Rassenkreis* (literally "race circle") and *Art* ("species" strictly speaking). A *Rassenkreis* is a genetical species with a series of intergrading but distinguishable local populations, occasionally so different that two terminal populations cannot interbreed directly even though still exchanging genes through intermediate populations. An *Art* is a genetical species without geographic differentiation and thus more nearly resembling the typological concept of a species. Mayr (1942) has, however, pointed out that most species are *Rassenkreise* and that although description of amount and kind of geographical variation is of great importance, a theoretical distinction between *Rassenkreis* and *Art* is not particularly meaningful in evolutionary taxonomy. Most modern taxonomists seem to agree (for example, Cain, 1954).

Along still another line, Mayr (1942) distinguished *allopatric species,* those occurring in different areas, and *sympatric species,* those occurring, at least in part, at the same localities. The genetical significance is that allopatric species do not actually have an opportunity to interbreed. They may even retain the potential for interbreeding while not in fact interbreeding and while undergoing considerable (but not indefinitely great) morphological and functional divergence. Application of the genetical definition to them may thus become blurred and uncertain. Application of the evolutionary definition is much clearer in such cases, at least in theory, although a practical question may arise as to whether the geographical separation entails a separation in roles. Sympatric species ordinarily remain such only if they interbreed within and not between species and do maintain quite separate roles. Both definitions usually apply unequivocally to them and lead to recognition of the same groups as species.

SPECIES IN UNIPARENTAL ORGANISMS

As already noted, the usual genetical definition of the species is relevant only for biparental organisms. Most animal taxonomists (and animal geneticists) have tended to ignore uniparental organisms and to make no attempt to define the category species with reference to them. There is some justification for that attitude, because long-continued obligate uniparentalism is rare among metazoans and the practical problem arises mainly in protists and plants. Even if biparentalism occurs alternately or at rather long intervals with intervening uniparental generations, the genetical concepts and definitions are usually still relevant. Moreover, Pontecorvo (1959) has pointed out that some form of biparentalism, whether sexual, or, in his term, parasexual,[10] is present in some members of practically all large groups of organisms, including protists and plants, and raises the probability that *all* organisms now living had either sexual or parasexual, therefore biparental, ancestors.

It remains true that there are obligate uniparental organisms, including some animals, and that they must be classified. It is further true that students of those animals recognize among them groups that seem to be the same sort of thing as species in biparental animals and that are also called species and named as such. They obviously cannot be genetical species by the accepted definition, and in fact Dobzhansky (1937) has insisted that there cannot be a species category for uniparental organisms. This misapprehension, as I believe it to be,

[10] Reproduction is fully sexual if it involves meiosis (reduction of chromosome number antecedent to karyogamy) and karyogamy (union of sets of chromosomes from two individuals in a nucleus of one individual). Parasexual reproduction, recognized only in recent years, involves exchange of genic material between separate individuals in any of several different ways without meiosis and karyogamy.

is clarified by recognition of an evolutionary species category, as defined above, that is fully congruent with the genetical (biparental) species and includes it as a special case but is also fully applicable to uniparental species as another special case.[11] Meglitsch (1954), in an exceptionally clear and penetrating study, has reached essentially the same evolutionary concept of the species. He demonstrates not only that this concept does apply equally to biparental and uniparental organisms but also that the genetic implications for evolution are comparable in the two even though the genetic processes are different. His definition is as follows:

The species, in the case of uniparental and biparental organisms, may be visualized as a natural population, evolving as a unit in actuality, or retaining the capacity to evolve as a unit if artificial barriers are removed.

(I take it that "artificial" is meant to designate barriers extraneous to the organisms themselves, not "manmade.")

Evolution as a unit, or the retention of a unitary role, is maintained by community of inheritance, by the capacity for genes to spread throughout the population (which therefore has a gene pool), and by the inhibition of their spread to other populations. All three factors occur in both biparental and uniparental populations. For community of inheritance that is obvious, but it is less so for the other two factors. In both cases the spread of genes occurs by differential reproduction, genes increasing if their possessors have more offspring and decreasing or eventually disappearing if they have fewer. In other words, the role is determined and is circumscribed by natural selection equally in the two cases. Biparental reproduction produces recombinations of genes, increases variability, and makes the species more adaptable, but it has no decisive bearing on gene spread within the population.

In biparental organisms the spread of genes between populations involves interbreeding, which of course does not occur in uniparental organisms. But it also involves the movement of individuals from one

[11] When I proposed the evolutionary definition (Simpson, 1951) I remarked that "emphasis on unitary evolutionary role may even resolve the theoretical difficulty of defining species in asexually reproducing groups." However, I stupidly prevented that application by including the word "interbreeding" parenthetically in the definition. In the meantime, and without knowledge of my passing remark, Meglitsch has cleared up the whole situation and I have reworded my definition appropriately.

population to the other, production of offspring there, and their integration and continued reproduction within the second population—all factors equally relevant to uniparental populations. Isolation in both is promoted by failure to migrate, by lack of effective reproduction in the new environment, or by failure of integration of the offspring. Biparentalism merely adds the factors that effective reproduction may be reduced by mating preferences and by hybrid sterility as well as by the extrinsic factors that affect both kinds of populations.

The evolution of uniparental and biparental populations is different in many important ways. That does not alter the fact that both form species and, by appropriate definition, the same kind of species.

SPECIES IN PALEONTOLOGY

The data for classification of fossils are characteristically different from those in neontology and present many special problems. The available anatomical data are always incomplete, and few or usually no direct physiological or behavioral data are available. Ecological data, although available in significant amount, are generally also much less complete than for recent animals. Large samples are less often obtainable and, incidentally, may be prohibitively expensive. The samples may be biased in numerous and sometimes peculiar ways. Paleontologists are in general well aware of these problems and are actively and, on the whole, successfully seeking workable solutions to them. I have summarized some of the most important of them elsewhere (Simpson, 1960a). They affect the application to specific cases rather than the general nature of taxonomic principles and I shall not go into them in this book.

As regards species, the paleontological approach frequently does not differ in principle from the neontological. Many available samples of fossils have no appreciable time dimension, that is, any span of time they may cover is too short to have involved determinable somatic change. In those numerous cases, the specimens studied may be considered contemporaneous, and the definition of the species category and recognition of specific taxa proceed on the same lines as for recent animals. A special point of principle, unique to paleontology, arises in the likewise numerous cases of temporally successive, appreciably changing samples from a single lineage or closely related lineages. The distinction of species in a phylogenetic pattern in time is related to their distinction in contemporaneous organisms but it presents some quite different questions. Those questions have also been discussed repeatedly and at length, for example, by Simpson (1951), Cain

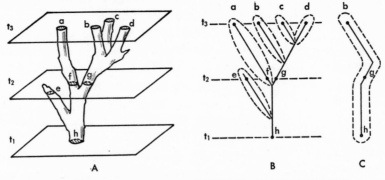

FIGURE 12. DIAGRAMS OF A PHYLOGENY AND OF GENETICAL
SPECIES AND LINEAGES WITHIN IT

A. Three-dimensional representation of a phylogeny. The horizontal dimensions within each branch symbolize extent of variation, and horizontal distances between branches represent degree of divergence between isolated lineages and species. The tree is represented as cut by time planes at times t_1, t_2, and t_3, and the intersects of each are genetical contemporaneous species.

B. Formal representation of the same phylogeny in two dimensions, with the same time intersects and genetical species (heavy dots). Each line leading to a terminal genetical species is a clearly separate lineage, enclosed in broken lines. The problems of classifying common stems and of possible successional divisions of longer lineages (e.g., *f–a*) require further consideration.

C. Lineage pattern starting from a terminal species, *b* as an example, and tracing back to the initial species, *h*, of the phylogeny. Each terminal species has such an extended lineage, and all overlap eventually as they are followed backward in time.

(1954), Sylvester-Bradley (1956, introduction by the editor, and consult the whole volume), Rhodes (1956), Westoll (1956), Newell (1956), George (1956), Imbrie (1957), and Grabert (1959), who also collectively provide references to much of the other pertinent literature.

In a branching phylogeny (Figure 12 A, B), the concept of the evolutionary species evidently applies to each separate branch. Cross sections make each branch that is cut a species of contemporaneous animals, generally corresponding with a genetical species. Each individual sample, or any sample that is not placed in a temporal sequence, can in fact be used for definition of an evolutionary species, or, at least approximately, of a genetical species if the assumption of biparental reproduction is warranted. In following the pattern through longer periods of time, however, even the evolutionary concept begins to run into trouble of at least two kinds.

An evolutionary species is defined as a separate lineage (also sometimes called a *gens* in paleontological usage) of unitary role. If you start at any point in the sequence and follow the line backward through time, there is no place where the definition ceases to apply. You never leave an uninterrupted, separate, unitary lineage and therefore never leave the species with which you started unless some other criterion of definition can be brought in (Figure 12C). If the fossil record were complete, you could start with man and run back to a protist still in the species *Homo sapiens*. Such classification is manifestly both useless and somehow wrong in principle. Certainly the lineage must be chopped into segments for purposes of classification, and this must be done arbitrarily (as defined in Chapter 4), because there is no non-arbitrary way to subdivide a continuous line. That is simple enough and fully justified on the taxonomic principles already expounded. In practice all that is needed is some criterion as to how large (and in what sense of "large") to make the segments. The following criterion is sensible and is accepted by almost all evolutionary classifiers of fossils. (All paleontologists are of course aware of the fact of evolution, but not all are evolutionary as taxonomists.)

Successive species should be so defined as to make the morphological difference between them at least as great as sequential differences among contemporaneous species of the same group or closely allied groups. (See Figure 13A).

Application of the criterion depends heavily on personal judgment and again reminds us that classification is an art. Nevertheless this is an artistic canon that limits the play of fantasy and provides for some guidance and uniformity. Adapting the previously stated range criterion for contemporaneous species, the rule may be made somewhat more readily and objectively applicable in this form:

When in a lineage the inferred ranges of observed changing characters for populations at two times do not overlap, those populations may be placed in different successive species and the division between species drawn approximately midway (in time) between them. (See Figure 13B.)

That criterion places the defining populations approximately in the middle of the segments that are designated as successive species. Just which samples are taken as representatives of the populations given this role depends largely in usual practice on chances of discovery and previous history of classification in the group. If, as frequently happens, one or more species are found and defined before the more extensive lineage is at hand, those species should be preserved as far as possible. That brings up a partial exception to the generally true statement that a type has no bearing on taxonomy except in nomenclature. The sample chosen as a datum for species-midpoint should include the type of any specific name previously proposed in that general part of the lineage. Types do in such cases help to determine just where along a lineage the boundaries of successive species are placed (Figure 13C). (They still have no relationship to the concept of species, and do not typify it.) Types have a similar centering effect for purely arbitrary subspecies, both contemporaneous and successional. (See additional discussion at end of this chapter.)

It has already been noted that segmental species in a lineage (or gens) are sometimes called paleospecies (or by the British, palaeospecies). I prefer to call them successive or, with Imbrie (1957), successional species. (Incidentally, Imbrie calls species among contemporaneous organisms, fossil or recent, *transient species*.) They are a distinctly different kind of thing from a genetical or other contemporaneous species, but I now think that the difference has sometimes been overstressed, by me as well as by others. They can both be viewed as aspects or states of the evolutionary species: one is a segment of an evolutionary species delimited in a certain span of time; the other is a cross section of an evolutionary species at any one time.

The other main problem of principle that arises in applying the evo-

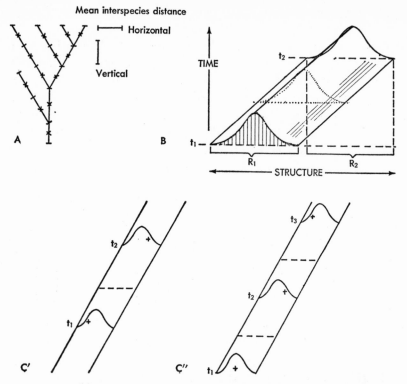

FIGURE 13. DIVISION OF LINEAGES INTO SUCCESSIONAL SPECIES

A. Criterion of approximately equal horizontal and vertical differences between species in a phylogeny like that of Figure 12. The lines follow approximate midpoints along the lineages of varying populations. The approximate mean horizontal difference between species can be used as a yardstick to divide the lineages successionally, as shown by crossbars. The crosses indicate approximate midpoints, in time, or centering populations of the successional species.

B. Criterion of range overlap in a single changing lineage. Population variation is symbolized by a normal curve, at any one time, generating a solid figure as structure changes through time. At time t_2 the range R_2 is distinctly outside that of R_1 at time t_1. The corresponding populations are placed in different species, divided arbitrarily at an intermediate time in a population indicated by the dotted curve.

C. Effect of types on subdivision of a lineage into successive species. In C′ types, the crosses, had been made of specimens in populations at times t_1 and t_2. The lineage is divided into two species at the time indicated by the broken line. In C″ three types had been made in populations at t_1, t_2, and t_3. Three successive species (or subspecies) are separated at times indicated by the broken lines.

lutionary species concept to fossil sequences has to do with branching. If the sequence reveals two lineages and their common stem, that stem, which cannot belong to both descendant species, may be placed in one of the latter, in the other, or in neither (Figure 14). In principle, the best solution is available when the lineages can be divided into three species, one ancestral and two descendant, separated at the point of branching. If, however, one branch changes much less than the other from the ancestry, either because it is much shorter or especially be-

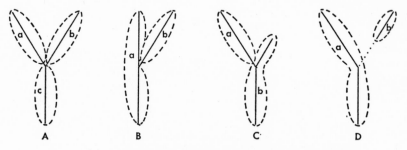

FIGURE 14. ALTERNATIVE SPECIFIC SUBDIVISIONS OF SHORT, BRANCHING LINEAGES

A. Three species separated at the point of branching.

B. Terminal *a* postulated as like the ancestral stem and included in the same species, while more divergent *b* is made a separate species.

C. Terminal *b* postulated as short and therefore little divergent from the ancestral stem, with which it is classified, while *a* becomes more divergent and is made a separate species.

D. A gap (dots) due to nondiscovery used for nonarbitrary division of species.

cause it is more conservative (evolves less rapidly), then that branch may well be placed in the same species as the ancestral stem.

Such situations do arise in paleontological classification, but as a matter of fact they are infrequent. Essentially continuous sequences of samples including the actual branching of a lineage into two or more are, as yet, comparatively uncommon. One reason for that is that speciation is usually allopatric. Two descendant species usually arise in different geographic areas within the range of the ancestral species, so that successive samples from any one place cannot show the branching, and a few geographically localized and scattered samples, such as are usual in paleontology, are likely to miss it. As previously explained, such gaps permit delimitation of species that is nonarbitrary as far as the data in hand are concerned (Figure 14D).

Other problems, not so much of principles as of their applications, arise from the difficulty, sometimes the impossibility, of determining the true temporal sequence of populations represented by paleontological samples. Stratigraphic correlation is still so imprecise that the relative ages of samples of approximately the same age but from different localities often cannot be determined. It can also happen that a

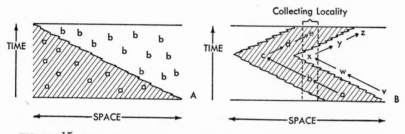

FIGURE 15. SOME PROBLEMS OF SAMPLING GEOGRAPHICALLY SHIFTING PALEONTOLOGICAL SPECIES

A. Two species, *a* and *b*, each confined to an environment shifting geographically. At any one locality, *b* is stratigraphically above *a*, hence later, but in their whole distribution the two species are contemporaneous.

B. Two related lineages, *a–e* and *v–z*, evolving in different, geographically shifting environments. Neither lineage can be followed at any one collecting locality. Fossils at the collecting locality designated by way of example give the temporal sequence *b–x–e*, which is not a lineage.

species always found above another, hence later in age, at each single locality is really contemporaneous with the other (Figure 15A). Shifting geographic ranges of related populations or species introduce many such problems and apparent anomalies. (Another example in Figure 15B.)

Two samples of approximately the same age but from different localities are never exactly the same. If the difference is acceptably significant in a statistical sense, it may still be due to geographic variation or to temporal evolution, or to a combination of the two. Intervening samples if available (and in paleontology they often are not) will show whether there was continuity between the populations but may not clearly indicate whether the continuity was geographical in one genetic species or temporal in an evolutionary lineage. In such very frequent instances, with or without intervening samples, the previously given canons of range overlap or separation can only be arbitrarily applied. The classifier is then recognizing species by a reasonable criterion, but

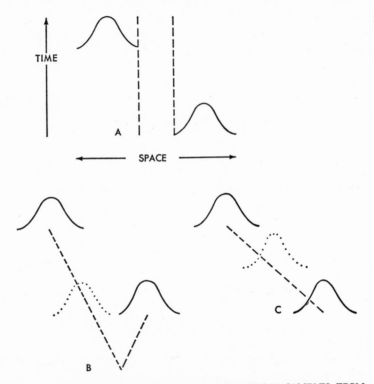

FIGURE 16. INTERPRETATION OF PALEONTOLOGICAL SAMPLES FROM SOMEWHAT DIFFERENT HORIZONS AND LOCALITIES

A. The given situation, with variation of inferred populations symbolized by curves. There is no range overlap (gap shown by broken lines), and the populations would ordinarily be placed in different species.

B. Interpretation as two lineages (symbolized by broken lines) of common ancestry. Discovery of such a population as that shown by the dotted curve would strengthen this interpretation.

C. Interpretation as successive species of one lineage. Discovery of the population shown by the dotted curve would strengthen this interpretation. (Still a third possibility is shown in Figure 17.)

he can rarely be quite sure as to just what kind of species they are, particularly whether they are successional or belong to different contemporaneous lineages (Figure 16).

SUBSPECIES

It is among the truly basic and universal facts of nature that all species vary. In time, they vary: cyclically—by days, by seasons, by years; sporadically—in irregular episodes that are unusual or unique; and secularly—in the long trends of evolutionary progression. In space (at any one time), they vary: within local populations (or demes, below), and geographically between nonisolated, incompletely isolated, or temporarily isolated local populations. Most pre-evolutionary taxonomists could ignore these facts, so inconvenient for the prevailing typology. For temporal variation they had practically no data beyond those of short (for example, annual) cycles that simply returned the populations to a previous state, or seemed to do so. For spatial variation their small and, as a rule, widely scattered geographic samples could be, and in great numbers were, considered representative of distinct species if they differed appreciably.

Even before Darwin, increased knowledge of geographic distribution and variation began to make it evident that populations completely intergrading, hence of one species if species are really separate units, nevertheless frequently may be definably different in different parts of the area covered by the whole species. As early as 1844 ornithologists began to recognize such geographical groups as taxa in classification and to give them trinomials, adding a third name after the Linnaean specific binomial (apparently first by Schlegel; see Sibley, 1954). There was much immediate opposition to that innovation, and objections to the subspecies as a distinct taxon and to trinomials have continued sporadically ever since.[12]

[12] A recent flare-up of the controversy followed Wilson and Brown (1953), and the journal *Systematic Zoology* has subsequently devoted much space to numerous, sometimes repetitious and somewhat inconclusive discussions of this point. It is unnecessary for my more general purposes to cite or review all those articles, or all those stimulated by them but published elsewhere, and the files of *Systematic Zoology* may be consulted seriatim by those especially interested. Smith and White (1956) may be cited, by way of example, as one of the most complete replies to Wilson and Brown.

Early objections were mainly typological, although usually not overtly so and almost never put in just those terms. If two populations are distinguishable, they have different archetypes. The embodiments of archetypes of lowest categorical level are specific taxa. Therefore the subspecies of later authors are "really" species, and should be so named as they would have been if recognized in 1758, or else they are no taxa at all but only accidents within species. Even in fairly recent years, that has still been operationally, at least, if not philosophically the attitude of a few taxonomists (for example, Kinsey, 1936; Burma, 1948) who insist that the smallest demonstrable differences in populations qualify them as separate species. That attitude over the last century or more (commonly associated with a "splitter" personality) has, incidentally, resulted in the designation of many thousands of "species" that most modern taxonomists consider subspecies, at best, if not entirely invalid as taxa.

The vast majority of modern taxonomists agree that species are not definable as minimal "types" but in genetical and evolutionary terms, and that when so defined they are usually "polytypic"—an almost universally used term that is nevertheless unfortunate because it embalms the discarded concept. Persisting disagreement among truly evolutionary taxonomists is therefore not over the question whether species have distinguishable geographic subgroups—most of them obviously do—but about the practicability and advisability of recognizing such subgroups as taxa and naming them (as trinomials) in the usual formal classifications. All the arguments really boil down to two questions:

1. Is there any nonarbitrary element (or "objective basis") in subspecies?

2. Is this recognition of a formal subspecific category (whether wholly arbitrary or not) useful to taxonomists?

The answer to both questions is, "Yes and no," or, "Sometimes." In some species there are nearly uniform populations over fairly large geographic areas separated from adjacent, more distinctive populations along lines that are narrow zones of intergradation. Those zones, as mentioned in Chapter 4, do divide species into relatively but not absolutely nonarbitrary subgroups. Those subgroups are commonly recognized as subspecies and they are "real" in that they correspond with groupings in nature and not solely in the taxonomist's mind. In many other species, however, no such relatively nonarbitrary or semiarbitrary subgroups are evident.

Some species show more or less steady changes in characters in one or more geographic directions: clines. It has frequently been suggested that clinal rather than subspecific designations then be used, but clines do not define taxa, as will be indicated in the next section. Other species show extremely irregular geographic variation, not organized in clines and not definitive of extensive subgroups. (Still others, rarely, show little or no appreciable geographic variation, but for them the question of subspecies does not arise.) Examples of all these situations have been given extensively by Mayr (1942, and much of his other work; see also his brief but incisive remarks on the subspecies polemics in Mayr, 1954).

When there are semiarbitrary subgroups in a species, their designation as subspecies should hardly raise any question if the data are adequate and the taxonomist finds the subspecies valuable for his purposes. It is no argument against such usage that recognizable subgroups do not occur in all species, because the subspecies is a nonobligate category and need not, even in principle, be used throughout the whole of a classification.

When geographic variation is clinal or irregular, any subspecies recognized must be wholly arbitrary. That again is not, in itself, an argument against subspecies, because many of the higher taxa universally and necessarily recognized are equally arbitrary. Nonarbitrary classification is flatly impossible. As to whether they are useful, the conclusive answer is that most of the experienced specialists in many groups of animals (notably the higher vertebrates) find them so and regularly use them. They can certainly be abused, but so can, and on occasion is, any category. The single species of pocket gophers *Thomomys umbrinus* has 213 currently recognized and named "subspecies" in southwestern United States and northern Mexico, and those who enjoy that game may well go on until every little colony of those gophers sports its own Linnaean name. There are nevertheless reasonable canons of the art, the application of which curbs such excesses. Most important are statistical methods that assure that at least a minimum proportion, now usually set at 75 per cent, of individuals of adjacent subspecies

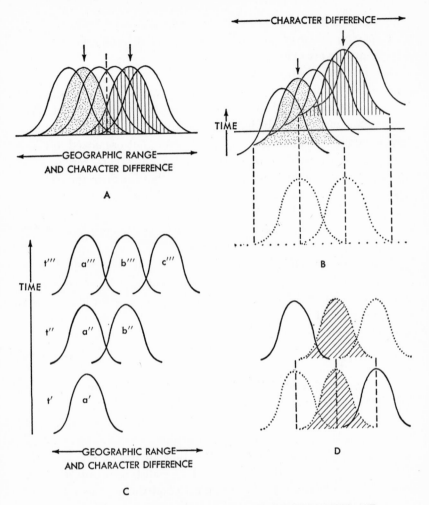

FIGURE 17. SOME RELATIONSHIPS OF CONTEMPORANEOUS AND
SUCCESSIONAL SUBSPECIES

A. A series of intergrading contemporaneous demes, each represented by
a variation curve. The two shaded curves, with modes indicated by arrows,
approximate the 75 per cent rule for dissimiliarity and are placed in separate
subspecies, the division between which is placed geographically as shown
by the broken line.

B. A series of intergrading successional populations in one lineage. If
projected onto a common time base, or if time is ignored, the relationships
are as in A, and the two populations indicated by shading and by arrows

will be unequivocally determinable (Amadon, 1949). That assures markedly significant differences in mean characters of the subspecific populations (Figure 17A).

Two misapprehensions about subspecies must be explicitly corrected. First, they are not little species, as typologists have thought, nor are they usually incipient or near-species, as some evolutionary taxonomists have assumed. They are taxa of a markedly different kind from species, and relatively few of them will ever become species although some are, to be sure, approaching that status. Second, subspecies do not express the geographic variation of the characters of a species and are only partially descriptive of that variation. They are formal taxonomic population units, usually arbitrary, and cannot express or fully describe the variation in those populations any more than classification in general can express or fully describe phylogeny. They are not, for all that, any less useful in discussing variation.

Subspecies are not in general use in paleontology, and there seems to be no definite consensus on this point although their paleontological use is probably increasing. Geographical sampling of fossils is usually inadequate for consistent recognition of subspecies of exactly the same kind as in recent animals, although a number of such cases are in hand and the number is increasing. Temporal subspecies, subdivisions of successional species, are different in kind from geographic subspecies. They can nevertheless be defined by similar procedures, such as the "75 per cent rule" mentioned above, and when this is possible they are extremely useful in stratigraphic study. The only likely alternative is use of unacceptably small successional species, and on this ac-

---

FIGURE 17. (*Continued*)

meet the 75 per cent criterion. They are placed in separate successional subspecies, separated at the *time* indicated by the horizontal line.

C. Geographic expansion and morphological divergence through time in a single lineage, in such a way as to make geographic and successional subspecies virtually equivalent. The successional subspecies $a'-b''-c'''$ are practically indistinguishable from the geographic subspecies $a'''-b''-c'''$.

D. A possible interpretation of the populations of Figure 16 as subspecies rather than species. Discovery of either or both of the shaded populations would suggest this interpretation, by which only the geographic and not the temporal difference between the two samples is pertinent to their dissimilarity.

count I feel that the use of successional subspecies should be encouraged. Some relationships and contrasts between contemporaneous and successional subspecies are shown in Figure 17.

Use of the concept, although not the term, is old in paleontology, going back at least to Waagen (1869), who applied to successional subspecies the term "mutations" before the geneticists captured that word. Bather (1927) substituted the term "transient" (a usage different from that of Imbrie, 1957, previously noted). (See Rhodes, 1956.) The term "waagenon" has also been proposed (by Caster and Elias; see Newell, 1947) and is probably least open to misunderstanding if a terminological distinction from geographical subspecies is desired. The principal argument for calling waagenons successional subspecies is that this gives them an appropriate place in the regular hierarchy of classification.

OTHER INFRASPECIFIC GROUPS

Neither the species nor the subspecies is the ultimate systematic unit of populations in nature. The smallest unit of population that has evolutionary significance is a group of individual animals (of one species or subspecies) so localized that they are in easy and more or less frequent contact with each other, the unispecific members of a single community in the most limited sense. That minimal population unit is called a *deme*. In biparental animals, the chances that a mate will be found within such a group are much greater than for matings between such groups.[13] In both biparental and uniparental populations this is the group within which infraspecific competition and selection tend to be concentrated.

A deme may correspond with a subspecies or even, in extremely exceptional cases, with a whole species. It is, however, almost always a decidedly smaller group than either of those taxa. Some demes are quite

[13] For that reason, Dobzhansky (e.g., 1951) and some other geneticists call biparental demes *Mendelian populations*. I prefer the term deme because it does not have the unnecessary limitation to biparental organisms, and also because the use of "Mendelian" in this sense is open to some slight question. Before the term deme came into general use taxonomists often used the vaguer and clumsier expression *local population* in the same sense.

clear-cut: populations on small islands, in separate groves of trees, in anthills, and the like. More often, they have vague or really no boundaries and depend on the point of reference. In an evenly distributed specific population the deme pertinent to individuals is different for each of them, as it covers the other individuals that each regularly encounters.

Even though demes are the basic population units, they do not and should not enter into classification as such and should not be named. That is not because they could usually be defined only arbitrarily—such is the case with most taxa, as we have seen. Demes are not classified, first of all, because they do not have long-continuing evolutionary roles and are highly evanescent. In subspecies, too, the differentiation of roles within the unitary specific role is incomplete, and subspecies are also evanescent, but there is some appreciable differentiation and the groups do ordinarily last long enough to seem stable in terms of human history. Adjacent demes frequently have no observable differentiation, and they may merge, split, or fade away during the lifetime of one taxonomist. Such groups are highly instructive from an evolutionary point of view, but they do not lend themselves to formal, even roughly stable classification. Demes are also so extremely numerous that no one could cope with an array of names for all of them, even within, say, one genus of abundant animals. (The example of "subspecies" in pocket gophers previously mentioned approaches a classification and naming of demes, but to the extent that it does so it is bad classification, a confusion of the concepts of subspecies and demes, and not a demonstration that demes, as such, are acceptable taxa.)

One of the commonest and most abused terms in taxonomy has been *variety*. On typological principles, each species had a fixed pattern. Anything that did not fit the idealized pattern was a "variety," a conceptual extension of the scholastic "accident." Like so many typological bequests, the term and concept were taken over by early evolutionary taxonomists and have continued to plague and confuse the science. It is rarely clear whether a variety is supposed to be: (1) an individual variant, (2) a group of such variants or morphs conceptually associated by the variation alone and not forming a population, or (3) a distinguishable population within a species analogous to or perhaps identical with a subspecies.[14] Early (and a few later) evolutionists not only failed to make those distinctions but also did not realize that the distinctions exist and are vital for taxonomy. That fault is probably the most serious logical ambiguity in the usually severely careful work of Darwin.

Varieties, usually without any clear or defined meaning, have frequently been inserted in classifications and given trinomial, quadrinomial, or some other kind of symbolic designation. There can be no serious doubt that the variety as a category in classification should now be abandoned altogether. (See, for example, Newell, 1947.) In the first two senses it does not refer to populations and therefore cannot have taxa, which are composed of populations by definition, as members. In the last sense, it refers to populations acceptable as taxa only when it is synonymous with subspecies.

Such terms as *form* or *morphotype* have been applied to varieties in the second sense, that is, to the aggregate of particular variations within populations rather than to all varying organisms forming a population. The terms are useful in describing variation without the ambiguity of the term variety, but they also obviously cannot designate categories of classification. Edwards (1954) has proposed that *morphs* be defined as "distinguishable sympatric and synchronic interbreeding populations of a single species," which he believes also to be subspecies of some authors and demes. In fact sympatric and synchronic members of a single species cannot form separate populations in the modern taxonomic sense but only "varieties" of the second sort, or forms or morphotypes, in short morphs, of a polymorphic species. They are neither subspecies by any now current definition nor demes. Edwards's injunction that they should not receive technical names as taxa can only be applauded.

A different sort of concept also applicable to variation within a species is represented by the term *cline*, proposed by Huxley (1938) and now in general use. Huxley's original, broad definition was: "A gradation in measurable characters." If not otherwise qualified, it is now

---

[14] The term "variety" is now most widely familiar in the sense taken over from botany into horticulture. Those usages also confuse the second and third senses of the word. The term and its ambiguity can hardly be eradicated from horticulture, but that need not concern the scientific animal taxonomist.

usually understood that the gradation of a cline is within a single species, that it is unidirectional, and that it is more or less uniform rather
than distinctly steplike. Many different kinds of clines have been
recognized, and a rather complex terminology was proposed by Huxley
and has since been supplemented: *geocline* for geographic, *ecocline*
for ecological, *chronocline* for successional clines, and many others.
The most frequent usage is, however, for geographic clines, and those
are now usually understood when cline, alone, is stated.

In his original proposal Huxley said that specification by clines was
intended to supplement and not replace current taxonomic methods.
Almost at once, however, he added (Huxley, 1939) that specification
by clines in certain instances should replace "specification by named
areal groups," that is, subspecies. The suggestion has been followed up
to some extent, but in general clines have not been adopted as the basis
for categories in classification or for zoological names. The clinal concept has proved to be applicable to almost all groups of animals and
to be extremely useful in representing and interpreting their variation.
It has become indispensable in systematics, but it is not a truly *taxonomic* concept as taxonomy is more sharply defined.

The point at issue here has been well expressed by Mayr (in Sibley,
1954):

Subspecies and clines are concepts belonging to different fields. A population can belong to only one subspecies but it can belong to several
different clines. In other words, the subspecies is a taxonomic concept while
the cline is an evolutionary one.[15] Nothing is gained by naming clines.

A cline is an arrangement of characters, not of organisms or of populations. It thus can happen, and very frequently does, that different
characters give quite different clines, running in different directions,
at right angles to each other, and so on, within the same species—a
fact that Huxley noted from the start and that is involved in the quotation from Mayr. In short, taxonomic classification is concerned with
the arrangement of populations and the cline concept is not. Therefore that concept cannot be used to define any taxonomic category.
It is, of course, extremely useful in arranging the data of classification

[15] Of course Mayr does not mean that taxonomy is nonevolutionary, but that
this concept belongs to the nontaxonomic part of evolutionary systematics as the
terms are here defined.

and in understanding and defining individual taxa, as are all studies of infraspecific variation. In present classification, however, the *only* acceptable infraspecific *category* is the subspecies.

SUPERSPECIES

The next category above the species that is provided for in the Rules of Nomenclature and that is in wide use is the subgenus. It is discussed, along with the genus and higher categories, in Chapter 6. There are, however, groupings occasionally used between species and subgenus that may appropriately be discussed here because they tend to merge rather with the polytypic species concept than with the genus.

Many large (highly polytypic) genera contain clusters of closely similar species. Monographers frequently find it convenient to designate these clusters and have done so under various terms, such as *species groups*. They are not formally assigned categorical rank in the hierarchy and do not have special names but are referred to by numbers or other symbols or, most commonly, by the name of an included species, for example, "the *Drosophila macrospina* group." In some instances in the literature the groups seem to be better definable as species and their included "species" as subspecies. Even more commonly they could and in my opinion should be given properly hierarchic designation as subgenera, a category that some taxonomists are curiously reluctant to use, as I have mentioned.

A category superspecies could be intercalated in the hierarchy in a normal way, just as superfamilies, for example, have been. Mayr (1931, 1942) has proposed that this term be applied to what was formerly called an Artenkreis, defined in Mayr, Linsley, and Usinger (1953) as follows:

A superspecies is a monophyletic group of very closely related and largely or entirely allopatric species.

Superspecies are, in other words, groups of populations that seem on other grounds (morphology, ecology, etc.) to have passed beyond the point of potential interbreeding and to have acquired separate evolutionary roles, but that are not demonstrated to have done so by the more conclusive evidence of remaining separate when sympatric. It is to be assumed that they are still near the critical point of speciation, that of definitive isolation, and it cannot be quite certain whether they are really past that point and are not just below it. They are nascent species that will, if they survive, collectively form a subgenus or eventually a genus but have hardly yet reached that degree of divergence and expansion. They are not given special names; the Rules of Nomenclature make no provision for that. Usual designation is by the name of an included species, as for species groups, some but not all of which are in fact superspecies *sensu* Mayr.

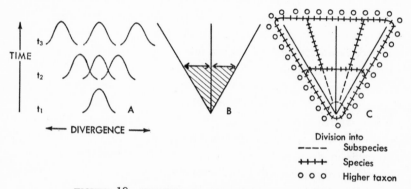

Division into
- - - -   Subspecies
++++   Species
o o o   Higher taxon

FIGURE 18. ANALYSIS OF SUPERSPECIES CONCEPTS

A. Successive and ancestral-descendant populations, symbolized by variation curves. At t_1 there is a monotypic species, at t_2 a species with three subspecies, and at t_3 three closely similar species. Superspecies of Mayr or geographical superspecies of Sylvester-Bradley designates the stage of speciation seen at t_3.

B. The lineage pattern of A. In the shaded portion, below the double-headed arrows, interbreeding continues and the lineages are not isolated. Each lineage (all three intergrading or overlapping below) is a chronological superspecies of Sylvester-Bradley.

C. Division of the indicated lineage pattern into temporal-spatial taxa.

Sylvester-Bradley (1951, 1954) has suggested that the term superspecies be applied not only to contemporaneous species, the application of Mayr's definition, as *geographical superspecies* but also extended to their ancestral species through a *chronological superspecies* in each lineage involved. The whole segment of phylogeny would then, in, Sylvester-Bradley's usage, be a superspecies without adjectival qualification. Cain (1953, 1955) has objected vigorously. The polemic is in part unnecessary semantic quibbling—not that attention to semantic minutiae is generally unnecessary—and neither author seems to me to have made the important issues entirely clear. Cain is clearly right that three different concepts are involved, and Sylvester-Bradley is right that all are aspects or parts of a single, larger taxonomic concept. The issues do warrant careful consideration because they illuminate several taxonomic principles.

The phylogenetic pattern involved is shown diagrammatically in Figure 18. At time t_1 (Figure 18A) there is a monotypic species, which at time t_2 becomes a polytypic species with three subspecies, and at time t_3 a superspecies *sensu* Mayr or geographical superspecies of Sylvester-Bradley. We thus have a first concept, that of taxonomic divergence, one stage of which is Mayr's superspecies. The second concept (Figure 18B) is that of lineal descent, with three lineages in the diagram, each of which is a chronological superspecies of Sylvester-Bradley. The species of t_1 and t_2 belongs to all three. The lineages or "chronological superspecies" cannot be made taxa unless a horizontal division is made between t_2 and t_3, and that in itself breaks up the "chronological superspecies," which, in short, simply are not definable *as taxa*. The third concept (Figure 18C) is the only strictly taxonomic one: hierarchic subdivision of the phylogenetic pattern, as shown by the group-in-group boundaries in the diagram.

In effect, then, what Sylvester-Bradley has proposed is that the term superspecies be used in a regular taxonomic way for a taxon that is a section of a branching phylogeny larger than a species but smaller than a genus (or, presumably, subgenus). That would seem to be the most logical usage, for the superspecies thus falls naturally into the hierarchic category sequence subspecies–species–superspecies–subgenus– and so on. However, Sylvester-Bradley also uses the same name for elements, stages, and lineages, such that in this particular pattern they cannot be readily made taxa in the hierarchic sequence. The distinction is clearly made by Sylvester-Bradley's qualifying adjectives, but the nominal designation of all three concepts as superspecies and the conflict with Mayr's original and more widely known definition is nevertheless quite confusing.

If a taxonomic category between species and subgenus were needed, which is questionable, and if it were clearly defined and provision made for nomenclature of its member taxa, which has not been done, then superspecies would be the proper name for that category. As it is, superspecies will ordinarily be understood *sensu* Mayr, which very usefully describes a stage in a particular evolutionary situation but is, at best, equivocal as definition of a taxonomic category. For the latter reason I would prefer to return to the original term Artenkreis, which is more clearly distinct from the system of hierarchic categories. (The fact that it is German and may be mispronounced seems no more troublesome than the fact that the term hierarchy is mispronounced Greek.)

TYPES AND HYPODIGMS

This chapter on lower categories is an appropriate place to emphasize and to expand moderately a point already mentioned several times: the essential principle that taxa in modern evolutionary taxonomy are based on samples, not on types. Types have traditionally served three functions: (1) as the sole or principal basis for the description and *definition* of taxa (primarily of species); (2) as a standard of comparison, approximation to which warrants *identification* of another specimen as belonging to the same taxon as the type; and (3) as a vehicle attached to a *name*, carrying that name with it wherever the type may go when taxa are subdivided or united.

On strictly typological grounds there is no reason in principle why the same single specimen should not serve all three purposes. In preevolutionary taxonomy it was universally assumed that a type can and does serve all three adequately, and indeed there was rarely any perception that the three are different. That typological concept was one of many carried over, almost unconsciously, into early evolutionary taxonomy, and it has hung on as one of the most difficult to eradicate. In fact in truly and completely evolutionary taxonomy the three functions assigned to types are completely incompatible. No one specimen can possibly fulfill them all properly, and failure to recognize that fact has led to the retention of basic ambiguity and cryptic typology in otherwise evolutionary taxonomy.

It has already been emphasized often enough that taxa are inherently variable and that attention to their variability is essential in their description and necessary in their practical definition. That naturally demands taking into account all available specimens and involves the principle that no one specimen referred to the taxon is, for these purposes, any more important or any more typical than any other. Some specimens are of course more nearly average than others as regards particular characters in the sense of being nearer the mean, although this is rarely true of all characters of one organism. The mere fact that a valid average is recognized means that *all* specimens have been taken into account and none especially weighted.

That procedure does not mean that description of single specimens has been abandoned, as is disapprovingly assumed in a startlingly typological note in prominent recent publication (Shenefelt, 1959). One practical way to discuss variation is to describe in detail one specimen, which need not and often should not be the nomenclatural type, and to note differences in other specimens in the course of that description. In accompanying statistical characterizations, however, all specimens are of course simultaneously and equally involved.

In identification, comparisons of individual specimens are of course made routinely, and not necessarily with the nomenclatural type, which is often far from average in pertinent diagnostic characters. Final decision as to conspecific status depends, however, not on nearness to any one specimen, type or other, but on falling within or outside of ranges of variation *inferred* for the whole taxon. Those ranges are inferred by statistical concepts, and preferably also procedures, from all the specimens previously placed in the taxon. Throughout both definition and identification, the specimens are considered as a sample from which characteristics of populations are inferred. If only one specimen is known, it is necessarily both the type and the sole basis of concrete description, but population inferences drawn from it, and not the specimen itself, remain the basis for comparison and identification. The usable sample will of course increase as more specimens are collected and will be different for different classifiers with access to different collections. The proper basis for definition and comparison thus changes and, as a rule, improves as time goes on. That basis is not a fixed type that can be designated once and for all in the original description.

In order to emphasize that role and to define it clearly and succinctly, in my first theoretical discussion of those points (Simpson, 1940) I proposed the term hypodigm, which may be defined as follows:

The hypodigm of a given taxonomist at a given time and for a given taxon consists of all the specimens personally known to him at that time, considered by him to be unequivocal members of the taxon, and used collectively as the sample on which his inferences as to the population are based.

The concept (which of course antedated my consideration of it and coining of the term) has come into general use, the term less so, although it is also now fairly widely adopted, mostly by paleontologists. The objection has been raised (for example, by Mayr, Linsley, and Usinger, 1943) that it is an unnecessary synonym of *material*. As long as the concept is used, the term for it is unimportant. I do, however, still feel that "hypodigm" (or any alternative, properly defined technical term) should be used in place of "material" for three main reasons, among others: (1) it is much more precisely defined than the vernacular word "material," even as used by taxonomists; (2) it focuses attention on the concept and assures that the procedure is in fact based on that principle; and (3) by providing a different, technically labeled rubric in formal proposals and revisions of taxa it tends to confine the rubric "type," always included, to its proper function of nomenclatural type.

Problems that arise in application of the hypodigm concept are those of statistical sampling in general. Some of them have been well discussed by Newell (1949b). Prominent among these is the difficulty of specifying a single homogeneous population and assuring that a sample is entirely drawn from it. It might be ideal for each sample to come from a single deme at a single time, a requirement rarely met for any but minimal samples either in recent or in fossil animals. Fortunately such stringent specifications are not really necessary for drawing valid statistical inferences and may even be a drawback. Many of the sampling problems are discussed in Simpson, Roe, and Lewontin (1960).

A later reviser may find it advisable either to restrict or to extend the original concept of a taxon. For that purpose he will need to know exactly whence in the temporal and geographical distribution of the taxon the individual parts of the original describer's hypodigm were derived. He will also need to know the temporal and spatial location of the type for nomenclatural purposes if he splits the former taxon. For those reasons, Newell (1949b) has suggested retention of the term *topotype* for those specimens in the hypodigm "that come from [more or less] *precisely* the same horizon as the holotype [that is, nomenclatural type or simply type]." That requirement is, however, met much more completely and without confusion of "type" concepts by simply specifying the exact horizon and locality of each member of the hypodigm, either in publication or in the records of the collection.

Newell also feels that paratypes should continue in use because, "The designation carries some prestige and is some assurance that the hypodigm will not be broken up for exchange purposes. It also stresses the published documentary value of the specimens." But paratypes do not have special documentary value, and the true value of *all* the specimens with documentary significance is stressed by designation in the hypodigm, proposed in part for that very reason. As to the first point, the authoritative proposal has been made that paratypes be selected as "duplicates" of the type ("holotype") *in order to exchange them*.

Attention has been drawn to the fact that types do have a further practical function, in the centering of wholly arbitrary successional species and subspecies both contemporaneous and successional. That function can now be explained a little more clearly, and it is still more nomenclatural than taxonomic in any other way. The arbitrary taxa are cut from an unbroken sequence of populations, either in space or time. The centering is really not on any individual but on some one population, more restricted temporally, spatially, or both, within the sequence. Selection for that function of a population that includes the type tends to stabilize nomenclature, and is useful for that and really no other purpose.

6

Higher Categories

Although the species is the most nearly fixed unit in taxonomic theory, even species intergrade in evolution with taxa above and below them. Usages of the expression "higher categories" differ, but we shall assume that they are supraspecific categories, with species defined as in Chapter 5. The genus has a somewhat special status and calls for some consideration separate from that of other higher categories.

Most of the important principles, concepts, and procedures involving higher categories have already been discussed among the principles of taxonomy in general in Chapters 2–4. Some need to be mentioned again and some to be further developed in this more special context. Especially in dealing with contemporaneous animals, criteria of discontinuity (gaps), diversity, and divergence are essential. More broadly, whole phylogenetic patterns must be considered. Analysis of complex patterns reveals a number of simpler, distinguishable relational elements generally present in them. That is especially true of patterns of diversification or radiation, which frequently but not always reveal the existence of relatively nonarbitrary (in that sense "real" or "objective") higher taxa and subtaxa within them. There are numerous special problems, such as the fact that a group may not be *uniformly* divisible into taxa at a given level. The interplay of the various pattern elements is most readily understood from concrete examples, here mostly drawn from the Mammalia.

Taxonomy deals with the recognition and formalization of relationships among organisms. All such relationships have evolutionary bases and origins. In recognition and emphasis of that fact, it is fitting to close this study of taxonomy with notice of some of the evolutionary factors that underly the patterns of taxonomic categories.

WHAT IS A HIGHER CATEGORY?

Probably the most active and certainly one of the most interesting fields of special study in systematics today is the evolutionary investigation of variation within single species or groups of closely related species. All such studies are of value to the taxonomist, but they become strictly taxonomic in themselves only when they concern the recognition and interpretation of taxa. Only one infraspecific category, the subspecies, is now commonly recognized, and it is not in universal use. The universally accepted lower or lowest category, the usual C_1 of formal classification, is the species. It does intergrade with categories above and, when subspecies are used, below it, but characteristically it is unique in being nonarbitrary in both inclusion and exclusion when based on contemporaneous animals. Categories above it differ, as a rule, by including more than one of those nonarbitrary units, hence having internal discontinuities among contemporaneous groups. They are therefore nonarbitrary as to inclusion, but their definition depends on criteria other than that of internal continuity. The supraspecific categories, all with that feature in common, are generally what is meant by higher categories. The distinction is not always clear, however, and the preceding discussion of lower categories (Chapter 5) may be supplemented by three examples of different understandings of the matter.

Merriam (1918) examined many skulls of North American brown bears and placed each individual that he found appreciably distinctive in a separate "species." Those, the basic units of his system, are not in fact and perhaps were not even in intention populations in any meaningful biological sense.[1] In fact they and the "genera" in which they were grouped are not taxa at all and belong to no category of proper modern classification.

---

[1] I no longer maintain a former opinion (Simpson, 1953) that Merriam tended to or was attempting to recognize and name demes, which are units of population even though below the level of practical taxa in classification.

Kinsey (1936), in one of the most substantial studies ostensibly devoted to higher categories, defined the species as "the lowest taxonomic unit recognizable among plants and animals," and considered any category above that level as "higher." His "species" are population units, but are not species by genetical or evolutionary definitions. Some seem to be demes and some might be considered subspecies, true taxa but of minimal rank. Kinsey's "complexes," between his "species" and his "subgenera," sometimes approximate genetical species. Very little of his monograph on "higher categories" has any pertinence to higher categories as here understood.

Scheffer's (1958) excellent book on the seals and their allies (Pinnipedia) deals, among other things, with higher categories by any definition, but it confuses the matter at levels of genus and species. Abbreviated and paraphrased, his criteria for genera are: (1) at least one variate does not overlap with other genera when individuals of the same age and sex are compared; (2) the way of life is distinctive, although not always evidently so for allopatric genera; (3) the breeding range is confined to one or two broad areas and may be sympatric with other genera; (4) different genera do not interbreed in the wild; and (5) they are usually distinguished by natives of the regions in which they live. Those are all characteristics (even so, not definitions) of the evolutionary *species*, not genus. (The last criterion given is unusual, but has some validity for species, as field workers know.) Scheffer's "genera" are thus true taxa, but not higher taxa.

At the other extreme, there is some tendency (especially among paleontologists) to consider the genus, rather than or in addition to the species, as the or a fundamental taxon. When they speak of higher taxa they usually mean families and above. The term "higher" is relative and inevitably has different connotations in different contexts. As will be noted, the genus also does have a special status. Nevertheless, the most important categorical break in principle is between species and genus, and for present purposes all supraspecific groups, with species defined as in Chapter 5, will be considered higher taxa. A higher category may be defined in terms of phylogenetic pattern as follows:

A higher category is one such that a member taxon includes either two or more separate (specific) lineages or a segment (gens) of a single lineage long enough to run through two or more successional species.

The second alternative accounts for the fact that higher taxa are often monotypic at any one time, such as the Recent, and hence that higher categories cannot be defined simply as those with taxa including two or more known species.

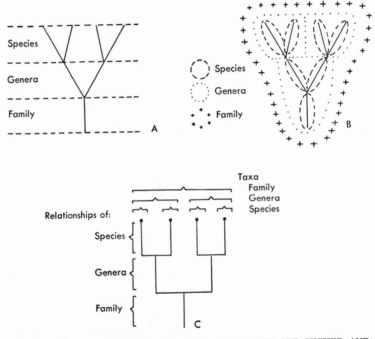

FIGURE 19. RELATIONSHIPS BETWEEN PHYLOGENY AND HIGHER AND LOWER TAXA

A. Phylogenetic tree with stems and branches incorrectly conceptualized as corresponding with taxa at different levels.

B. Same correctly subdivided into taxa.

C. Dendrogram showing the terminal species (black dots) *only,* and with lines below indicating neither phylogeny nor taxa but *relationships* among the terminal species, categorized as at left; the resulting *taxa,* with no time dimension, are indicated by braces above. As far as the terminal species alone are concerned, this is the same arrangement as that of temporal-spatial taxa in B, and it is correctly consistent with the inferred phylogeny, which is topologically equivalent to the relationship lines of C. (The diagram is artificially simplified and would represent undue splitting in most cases with so few species.)

This is an appropriate place for final clarification of a confused point that has been mentioned before: the relationship between a phylogenetic tree and higher or lower taxa. That relationship is often shown as in Figure 19A, as if the lower stems were, or represented, higher taxa and the upper stems and branches to the terminal species were, or represented, lower taxa. Such a conception is flatly false. The correct relationship of temporal-spatial taxa to the phylogeny is shown in Figure 19B, and the bearing on the classification of the terminal (for example, recent) species is as in Figure 19C.

BASES FOR RECOGNITION OF HIGHER TAXA

Most classifications are, in practice, based on contemporaneous organisms or have been made without explicit consideration of the time dimension. The criteria for higher categories as usually given reflect that limitation. The three principal criteria on that basis are: degrees of separation (gaps), amount of divergence, and multiplicity of lower taxa. Those will be considered first, and related to some of the phylogenetic patterns that produce them. The phylogenetic patterns, per se, of higher categories will later be further analyzed and exemplified.

Mayr, Linsley, and Usinger (1953) define all the higher categories by two points: monophyly and gaps. Their definitions for genus, family, and suprafamilial[2] categories in terms of member taxa may be combined and paraphrased as follows:

A T_{j+1} is a taxon including one or more T_js of inferred common phylogenetic origin and separated from other T_{j+1}s by a decided gap.

They further recommend that the size of the gap (that is, the degree of dissimilarity from adjacent T_{j+1}s) be inversely proportional to the size of (that is, the number of T_js in) the T_{j+1}. The general bearing and application of the definition are shown in Figure 20A.

This procedure's consistency with phylogeny depends in the main on two well-established and now familiar reciprocal principles:

1. Characters in common tend to be proportional to recency of common ancestry. The distances between lower taxa in this approach are inversely proportional to characters in common.

[2] Application of the definition to suprafamilial categories is not explicit but seems plainly implicit.

2. Degrees of divergence tend to be proportional to remoteness of common ancestry. Sizes of gaps thus tend to be directly proportional to remoteness of common ancestry among surviving lower taxa.

There is no doubt that these are usual tendencies and that, in the absence of other indications, the criterion does as a rule or on an average tend to produce consistency with the most *probable* inferential phylogeny. If the characters in common and degrees of divergence in Figure 20A have exactly followed the two principles, the phylogeny must have been topologically as in Figure 20B. As Michener (1957) has discussed at greater length and in somewhat different words, those average tendencies are not always followed in detail and the criterion may therefore produce small or, rarely, large inconsistencies with phylogeny. The phylogeny shown in Figure 20C is a distinctly less probable inference from the data of Figure 20A, but it is a possible origin for that pattern. If that were the true phylogeny, the genera shown in Figure 20A would still be entirely consistent with it, but in the more detailed arrangement the tribes and subfamilies would not be consistent. Figure 20D represents another, still less probable, phylogenetic pattern that would also but very exceptionally produce the results in Figure 20A. If that were the true phylogeny, even the genera of Figure 20A would be inconsistent with it.

No inferences are completely certain and no criterion, or artistic canon, always produces a uniquely valid classification. Apart from that point, the linear representation of Figure 20A is too simple, and if further realistic considerations were introduced further analysis would almost certainly permit a still more probable choice among different possible phylogenetic patterns. The one-dimensional approach involves reduction of degrees of similarity between any two populations or species to a single measurement, which may, however, subsume numerous different characters of the individual animals. That is essentially the result obtained by Sokal and Michener (1958). A still more extensive, but also still more laborious, approach through discriminants reduces numerous variates to two composite measurements and shows variation and gaps on a two-dimensional field. That is more nearly representative of the full comparison, which is multidimensional, but is still oversimple. Ashton, Healy, and Lipton (1957) have provided an elaborate example comparing *Homo*, australo-

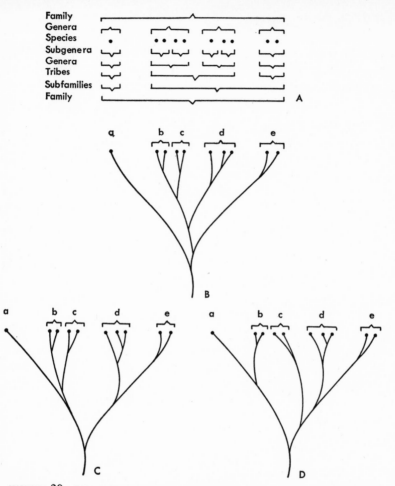

FIGURE 20. GAP AND MONOPHYLY CRITERIA FOR HIGHER CATEGORIES

A. Ten contemporaneous species (black dots), with gaps of various sizes, corresponding with degrees of dissimilarity, between them. Braces above show assignment of species to taxa of two successive higher categories, below to taxa of five successive higher categories.

B–D. Some possible phylogenies for the species shown in A. B. Most probable phylogeny, with propinquity of ancestry closely reflected in degrees of divergence; this is consistent with all taxa of A.

C. A less probable phylogeny, with less divergence from *c* to *d* than the remoteness of their descent would ordinarily produce; this is consistent with upper, not lower, taxa of A.

D. Still less probable phylogeny, with convergence between *b* and *c*; this is inconsistent with both upper and lower taxa of A.

pithecines, and apes, and they have shown that by choice of a plane of projection most, but not all, of the pertinent information can in fact be shown in two dimensions.

Ashton, Healy, and Lipton's results further show that for different anatomical parts (teeth in this case) and different dimensions the gaps revealed may be quite different in size and position. In several of their comparisons there are significant gaps between *Homo* and living apes, gaps long recognized on a qualitative basis and currently given familial rank. The australopithecines are in some respects on the *Homo* side of gaps and in others on the living ape, or recent pongid, side. Thus the australopithecines are neither *Homo* nor recent pongids (which was known in any case) but are a third group that does in some way, not directly evident from these data, tend to link those two groups. The data provide no evident direct basis for decision as to categorical rank of the australopithecine higher taxon, which might, for instance, be a third family, a subfamily of Pongidae or of Hominidae, or a genus in one or the other of those families.

Those and similar careful quantitative studies seem to demonstrate that measurable sizes of gaps are not in themselves adequate criteria for the ranking of higher categories and recognition of their relationships. Sokal and Michener did go on to form higher categories on the basis of their composite measurement of resemblance or "static relationships." Sokal (not primarily a taxonomist) apparently felt that this was the adequate and final result for taxonomy. Michener (who is a taxonomist) considered those results rather as helpfully analyzed and reduced data on which *phylogenetic* taxonomy could then be based in the light of other evolutionary considerations. (See also Michener and Sokal, 1957.)

On that point I agree entirely with Michener. Having measured or estimated degrees of resemblance and sizes of gaps, the taxonomist should then bring into the picture qualitative judgment as to homology, convergence, and the like. For example if Figure 20C were the true phylogeny, homologies between *a* and *b-c*, not present between *b-c* and *d*, would be highly probable and would indicate the monophyletic grouping *a-b-c* in spite of the large gap between *a* and *b*. Similarly in Figure 20D the convergent rather than homologous resemblance between b and c would, with high probability, be revealed by analysis with such possibilities in mind. The criterion or definition for higher categories involves monophyly as well as gaps (and still further criteria). Monophyly does tend to be related to the gaps, but it need not and should not be judged on that basis alone.

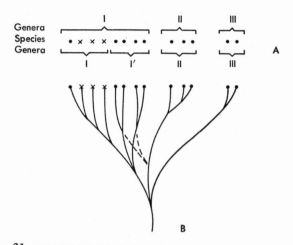

FIGURE 21. EFFECT OF FILLING A GAP IN THE GROUP OF SPECIES OF FIGURE 20

A. Species of Figure 20 plus three filling a gap (crosses); division into genera by gap criteria, alone, above, and by gap and divergence criteria below.

B. In solid lines, a probable phylogeny leading to the situation in A and consistent with both divisions into genera; broken lines show another possibility, consistent with the lower but not the upper division into genera in A.

Although phylogenetic divergence almost always does produce some gaps, it need not do so in principle and in fact does so only irregularly or inconsistently. Degree of divergence is in itself a useful and valid criterion for higher categories in addition to the criteria of monophyly and even in the absence of a definite gap. For example, Figure 21A shows the same species as Figure 20A plus three others that fill in the largest gap of the latter. By gaps alone, three genera would be recognized, but one has as much total divergence as the other two put together. It could, therefore, conveniently be divided into two genera, making four in all and of approximately the same size in terms of divergence. The four genera, although different in basis, correspond with those of Figure 20A. Such an arrangement would also be consistent with a probable phylogeny such as that of Figure 21B or, with addition of lineages, Figure 20B. In settling on an exact point for splitting the unduly large genus, indication of monophyly would of course again be taken into consideration.

An appropriate divergence criterion has already been suggested, and it may be restated as follows:

Within any one higher taxon, T_{j+2}, it is desirable that the higher taxa next below it in rank, T_{j+1}s, be approximately equally divergent in the absence of counterindications such as evidence of significant polyphyly of a T_{j+1} so formed.

The criterion of diversity, subsidiary to those already mentioned, has also been given earlier in a different context and needs only simple restatement here:

In the absence of counterindications related to monophyly, gaps, and divergence, it is desirable that the T_{j+1}s included in any one T_{j+2} include approximately equal numbers of T_js.

The point has further been sufficiently made heretofore that equal antiquity of common ancestries is *not* a desirable or even a possible criterion for placing higher categories in the same rank. Figure 22 may here serve as a reminder and further exemplification.

DEFINITIONS AND CHARACTERISTICS OF HIGHER CATEGORIES

Absolute definitions of higher categories are impossible. One of the unique characteristics of the species is that it can be and is defined without reference to any other category. Definitions of higher categories can only be relative to those of other categories, specifying relative ranks in the hierarchy and set relationships to taxa. Thus the family, for example, cannot be defined as such but only in its relationships to categories of next higher rank (ordinarily superfamily, suborder, or order) or of next lower rank (usually subfamily or genus). The most general definition of any higher category, C_j, is that it is a set of taxa (T_js) such that each T_j includes one or, usually, more T_{j-1}s and is included (with or without but usually with other T_js) in just one T_{j+1}.

For each higher category there are criteria or characterizations, not strictly definitions, that are also comparative and relate to the member taxa. The most general criteria are that each T_j shall be monophyletic, in at least the broader definition of monophyly, and that the member T_js of any one C_j shall be comparable. Thus in practice the working

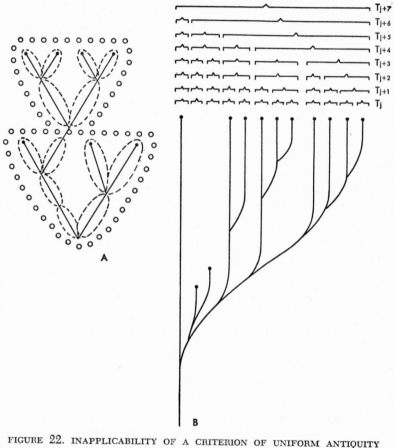

A

B

FIGURE 22. INAPPLICABILITY OF A CRITERION OF UNIFORM ANTIQUITY
OF COMMON ANCESTRY FOR TAXA OF THE SAME RANK

A. A simple phylogeny divided into taxa of two ranks, showing that a criterion of equal antiquity cannot be applied to a whole phylogenetic pattern.

B. Classification, indicated by braces above, of terminal members (black dots), only, by equality of antiquity of common ancestry, showing undue complication and imbalance of such a classification.

definition of a particular C_j, such as the family, is that it is a set of T_js, familial taxa, that are comparable among themselves and that are given family rank by the classifier who is doing the defining. That definition is necessarily subjective; so long as the hierarchic arrangement is maintained, ranks within it are assigned by art rather than by science. The vague word "comparable" is made somewhat more precise and the canons of the art are provided by special criteria, the most important of which are those of divergence, gaps, and diversity. Two other canons may here be added:

1. Between the species and a high category that has generally recognized and more or less stable member taxa, intermediate higher categories should be so spaced as to give approximately equal steps in the inclusiveness or extent and distinction of their member taxa. For example, the taxon Perissodactyla, now usually ranked as an order, has been almost completely stable since it was first recognized over a century ago. The recent species are enumerable without disagreement by those who apply the genetical definition. If, for example, the horses in the narrowest sense (one living species) were placed in one family and the zebras (three living species) in another, the steps from species to family would be exceedingly small in comparison with those remaining to be spaced from family to order. The horses and zebras should, by this canon, be at most distinct genera and (in my opinion) preferably subgenera.

2. When a high taxon is rather clearly separable into "natural" or semiarbitrary supraspecific groups, a definite category should be equated with their level, placed approximately according to the preceding canon, and other steps adjusted to that ranking. Among contemporaneous animals, this usually corresponds with a gap criterion because such groups almost always are separated by marked gaps at any one time. For instance, the stable taxon Carnivora is at present obviously and sharply divided into land carnivores (fissipeds) and aquatic carnivores (pinnipeds) at a level not lower than subordinal. Ranks from species to suborder are adjusted to that basic division. Some of the phylogenetic patterns involved in such a situation are discussed later in this chapter.

In current practice some higher categories do, even though very loosely and variably, tend to have special characteristics not wholly dependent on the preceding criteria and canons. It has long been felt and is rather deeply imbedded in the psychology of taxonomists that the genus is a different sort of category from those of still higher level. That was the opinion of Linnaeus, who crystallized it into all subsequent usage, even though his genus was a much higher category than ours. By still current Linnaean practice, the generic name is also part of the specific name so that one cannot name a species without naming its genus. (That is of course a long echo from scholastic usage in which genus and species *together* made a definition, at any categorical level.) The generic name is also unique in being an independent noun in the singular, whereas trivial names of species are either appositional or modifying, and suprageneric names are plural.[3]

The special status of the genus is not only traditional and nomenclatural. It frequently appears that the genus is a more usable and reliable unit for classification than the species. In dealing with classifications not erected or revised by modern and evolutionary standards— and many such must still be dealt with—it is often questionable whether its "species" are such by our definitions and not morphs or typological varieties, mutants, ecotypes, demes, subspecies, or something else. "Genera" are more likely to be acceptable taxa by modern definition, whether we would now assign them specific, generic, or some other rank. Even in modern usage, genera are often more clearly definable and defined than either infra– or suprageneric taxa. They are likely to lend themselves to more clear-cut characters-in-common diagnosis, lower categories, even species, being less clearly distinct and higher categories having fewer diagnostic characters in common and sometimes none. (This point is also made by Michener, 1957.)

Moreover in some kinds of investigation in systematics genera are frequently more appropriate or useful units than species or families. That usually stems from the fact that they are as a rule more widespread than species, especially in space but also in time, and yet are more conveniently localized and less nearly ubiquitous than many families. For example, in zoogeographic investigations on any but a purely local scale species are often too local and families too far-

[3] Cain (e.g., 1959) has several times suggested that the genus should not logically have special status in nomenclature and that binomials for species should be abandoned. Apart from the almost insuperable weight of tradition and hard-won codes, problems of homonymy and mnemonics would seem to make uninomials quite impractical for species.

ranging to bring out the significant features, while genera frequently turn out to be about right in those respects. In stratigraphic paleontology, species permit more precise local correlations, but genera are much more useful in long-range, especially intercontinental correlation. Also in paleontological studies of evolution, genera are commonly most useful because they are still reasonably small units but do have appreciable durations and are more largely and consistently sampled by fossil collecting. (Discussed in Simpson, 1960a.)

In dealing with fossils, among which alone the question arises in a direct and practical connection with classification, there is some tendency to use subfamilies and, perhaps to less extent, families as categories with long time dimensions, that is, as predominantly vertical units. There is great variation in this respect, however, and this can hardly be used as a rule or canon. A rather extreme example is Osborn's (1936, 1942) classification of the Proboscidea in which the 21 subfamilies have virtually no horizontal extent; each corresponds nearly with a single lineage. The family Equidae, as now defined, is a more balanced example, with considerable horizontal but still greater vertical extent. It also exemplifies a higher category that, like Darwin's Crustacea but for a different reason, practically cannot be diagnosed by characters in common. *Hyracotherium* (or eohippus) is much less like *Equus* than like many animals referred to other families, and *Hyracotherium* and *Equus* share no clear-cut character not also present in other families. Nevertheless *Hyracotherium* and *Equus* are adequately connected by sequential resemblances through a whole series of species and genera in time, very much as the Crustacea are by sequential resemblances in space.

As Mayr, Linsley, and Usinger (1953), among others, have noted, any one supraspecific taxon up to about the categorical level of families tends to have a fairly uniform adaptive facies, often recognizable to a taxonomist at a glance. Above the level of families there is "a basic structural pattern" and exceptionally there may be an obvious adaptive facies, as in the order Chiroptera (bats), but as a rule "the higher categories are not obviously or even predominantly distinguished by adaptive characters." The point has been touched on before and is again considered toward the end of this chapter. It may be otherwise expressed by saying that the common adaptive facies becomes more general and the basic adaptation is more likely to be overlaid by a diversity of others, or even to be lost in some members of the group, as one ascends the categorical scale.

ANALYSIS OF SOME PHYLOGENETIC PATTERNS

Numerous diagrammatic phylogenetic patterns have already been discussed and some of them figured in connection with various principles and procedures. It may now be useful to analyze them in a more systematic way and in their bearings on higher categories. This will be done (even here not in complete detail) by first abstracting the simpler phylogenetic relationships, then considering some combinations of the elemental relationships, and finally exemplifying these by real phylogenies authenticated by adequate paleontological evidence.

The most basic kinds of evolutionary events that build up real phylogenies seem to be only four in number:

Progression, or phyletic (that is, lineage) evolution; including but not quite synonymous with anagenesis of Huxley (1957).
Splitting or diversification; cladogenesis of Huxley.
Equilibrium, persistence or arrested evolution; stasigenesis of Huxley.
Termination or extinction.

It does seem desirable that these analytical factors have technical terms unequivocally defined, as Huxley (1957) has proposed. His terms "cladogenesis" (after Rensch, 1954) and "stasigenesis" are unequivocal. Unfortunately "anagenesis" is less so. Rensch (1954) defined it as parallel advance in organization or in perfection of function among the different lineages of a higher category. Huxley redefined it to include "all types or degrees of biological improvement," and when the term is now used there is a question whether it is *sensu* Rensch or *sensu* Huxley.[4] In any event, neither is quite what is wanted here, which is a term for *any* kind of change occurring sequentially in a single line of descent (or without reference to any branches that the line may also have), as symbolized in Figure 23A. Anagenesis *sensu*

---

[4] In fact the term had already been used long before Rensch in still other ways, but as those older usages have been practically forgotten they need not further obscure the matter.

Huxley is a special case of that more general concept. Huxley proposed no special term for extinction. Perhaps none is needed, but there is an ambiguity here, too, because taxa that died without issue and those that merely evolved so far as to be called different taxa of the same rank are both called "extinct." Termination probably more obviously designates the first, only, of those meanings. Termination can

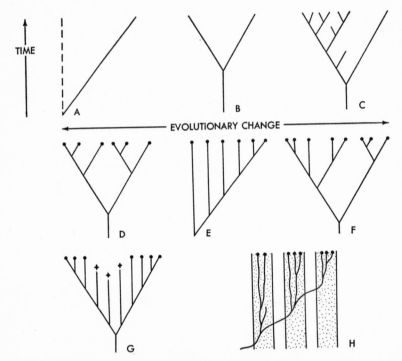

FIGURE 23. SOME ELEMENTAL PATTERNS IN PHYLOGENY

A. Simple progression; the broken line indicates the ancestral condition from which the lineage progressively departs.

B. Simple splitting and subsequent divergence.

C–E. Some of the simplest patterns of divergence and diversification. C. Unequal diversification of two diverging groups. D. Diversification producing a terminal array not sequential in origin. E. Diversification producing a terminal array that is sequential in origin.

F–H. Origin of gaps in terminal arrays. F. Gaps developing along with divergence. G. Gap caused by extinction of intermediate forms. H. Gaps between stable niches or adaptive zones (shaded).

also apply to the ending of a lineage in a group now extant, which is the same kind of phylogenetic *pattern* element, although of course a very different evolutionary situation.

The combination of progression and splitting gives rise to *divergence,* some of the simplest patterns of which are shown in Figure 23B–E. The examples given there do not exhaust the possibilities. It is also evident that these patterns are not sharply distinct but can grade one into another, and that they can be combined in many ways. Divergence also tends to produce the gaps so frequently helpful in recognizing and defining higher taxa, especially among contemporaneous groups (Figure 23F). Gaps may also be produced by prior termination of intermediate groups (Figure 23G), which of course may occur in any pattern of divergence and not only in the one exemplified. Gaps also arise by the occupation of separate levels of equilibrium, niches, adaptive zones, or grades (Figure 23H; the occupation is not invariably progressive as here shown by way of example). Those three modes of origin of gaps are not independent. The unequal divergence that forms gaps may do so precisely because its different lineages enter separate niches or adaptive zones, and the origin or maintenance of the separation of zones may be by extinction of lineages between them.

Progression, splitting, and accompanying divergence clearly tend to lead to increasing diversification or expansion, in evolutionary terms frequently an adaptive radiation, of what becomes through these processes a higher taxon. The simplest sort of expansion, certainly never quite so simple in nature, would be by continual dichotomy (Figure 24A). More realistically, the splitting may be figured as irregular and as accompanied by termination of some lineages (Figure 24B). The occurrence of extinction does not necessarily slow down diversification; the groups shown in Figure 24A and B are expanding at the same over-all rate. Another common variation of the simple pattern is for the main diversification to occur relatively rapidly near the base of the higher taxon with comparatively little over-all divergence thereafter and with later diversification of course at lower taxonomic levels (Figure 24C). That is a prominent element in rodent evolution, and it is exemplified over and over among widely different animals, for example, in many fossil invertebrates (for example, Nicol,

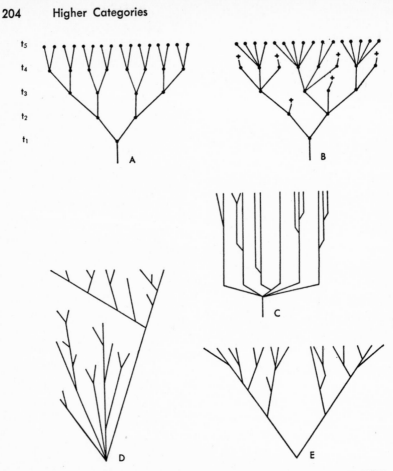

FIGURE 24. SOME PATTERN ELEMENTS IN EXPANSION OR RADIATION
OF HIGHER CATEGORIES

A. Simple, steady dichotomy, with doubling of lineages at regular time intervals, t_1–t_5.

B. Steady expansion at same over-all rate as in A, but with extinction of some lines and polytomy of others.

C. Rapid basic diversification followed by only minor diversification at lower taxonomic levels.

D. Successive radiations from the same origin, with the later radiation replacing the earlier.

E. Separate simultaneous radiations from the same origin.

Desborough, and Solliday, 1959).[5] Two other relatively simple variations, also common elements in known phylogenies, are the occurrence
of successive radiations, one of which may replace the other (Figure
24D) and contemporaneous but separate radiations of common origin
(Figure 24E).

It is the occurrence of divergence with gaps and more particularly
of separable radiations from the same stock that tend to produce
many of the semiarbitrary subdivisions of higher taxa as discussed
in Chapter 4. Four of the common phylogenetic patterns that have
that result are diagrammed in Figure 25 A–D. The grade pattern (for
example, Figure 23H) also generally has this taxonomic outcome. Many
phylogenies, despite the complications of more numerous lineages
and various expectable irregularities, can indeed be analyzed into
elements, and taxa, as simple as these patterns in principle. There
are nevertheless groups in which the patterns are so diverse that the
formation of *comparable* taxa of a given category may be practically
impossible. Figure 25E shows such a situation diagrammatically, and
it is involved in the problem of rodent suborders discussed later in
this chapter. Such difficult groups may foil efforts at even spacing
of successively higher categories.

SOME EXAMPLES FROM MAMMALIAN PHYLOGENY

The nature and origin of supraspecific taxa have been exemplified
more fully elsewhere (Simpson, 1959a). The rest of this chapter is
largely an abstract and slight revision of that paper, to which reference may be made for further details. It is advantageous to draw
these examples from the Mammalia, because here there is often comparatively good direct evidence on phylogeny. The mammals can
thus provide a series of well-founded paradigms as guides for interpretation of other groups with little or no known paleontological evidence.

The order Tubulidentata, family Orycteropodidae, and living genus
Orycteropus of aardvarks exemplify an almost irreducibly simple
hierarchic sequence of higher taxa diagnosed and ranked on the basis

[5] This is what I think their data show; some of their own conclusions seem to
me to be untenable.

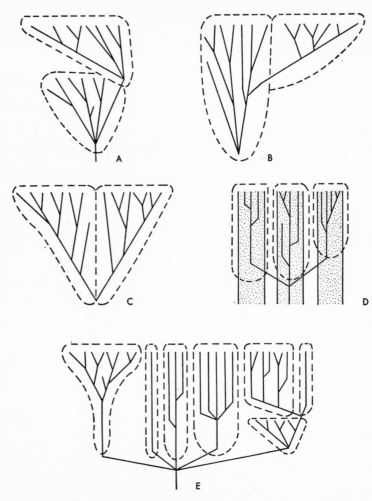

FIGURE 25. SOME PATTERNS WITH SEMIARBITRARY SUBDIVISIONS
OF A HIGHER CATEGORY

The subdivisions are enclosed in broken lines.

A. Subdivision by successive replacing radiation, as in Figure 24D.

B. Subdivision by successive radiations without replacement.

C. Subdivision by contemporaneous radiations, as in Figure 24E.

D. Subdivision by occupation of separate zones, comparable to Figure 23H but with zonal occupation not sequential.

E. Complex pattern with subdivision of such different kinds and scopes as hardly to be comparable or systematically placeable at a given taxonomic level.

of divergence and gaps. There is only one living species, so all the taxa named are monotypic in the recent fauna. The fossil evidence is not extensive, but from early Miocene, at least, onward it suggests no diversification above the generic level and perhaps, at any one time, none above the specific level. This extremely undiversified group is nevertheless classified as an order because its known members differ as much from any contemporaneous and possibly related orders as the latter differ among themselves. Differences from any possibly ancestral order are of the same order of magnitude. The gaps are, at present, absolute and are of the size now considered ordinal by almost all taxonomists. *Orycteropus* was formerly classified in the order Edentata because of resemblances that narrowed the gap from that group when viewed purely empirically. Those resemblances are, however, entirely convergent and must therefore be removed from any consideration of similarities and dissimilarities as a basis for evolutionary classification.

The basal gap between the Tubulidentata and their early Tertiary ancestry is doubtless due to nondiscovery. Hence this boundary as now drawn is nonarbitrary as regards the data but would almost certainly become arbitrary if the data were better. However, even if intermediates were found between the Tubulidentata and an ancestral order, ordinal ranking for the Tubulidentata would still be justified and an arbitrary horizontal basal boundary would have to be drawn. The terminal aardvarks differ to an ordinal degree from any possible earlier placental ancestry. The most probable ancestry seems to be the Condylarthra (see Colbert, 1941), a horizontally delimited order that more closely resembles its more surely linked ungulate descendants, notably the orders Perissodactyla and Artiodactyla. Practical classification demands some successional orders and arbitrary division between them, just as it demands some successional and arbitrarily defined families, genera, and species.

The family Leporidae, hares and rabbits, provides a comparatively simple example of a higher taxon with semiarbitrary subdivisions developed as more or less separate radiations. The phylogeny given in broad outlines by Dawson (1958) and shown in Figure 26 in a still simpler, diagrammatic way, suggests three separate radiations, partly successive-replacing and partly contemporaneous. From lineages of an early Tertiary radiation in Holarctica, two separate radiations developed, one apparently starting a little earlier and mostly Nearctic and the other initially Palearctic. These radiations replaced all the lineages not ancestral to them in the earlier radiation. There was subsequent interchange between Old and New Worlds in both the later radiations,

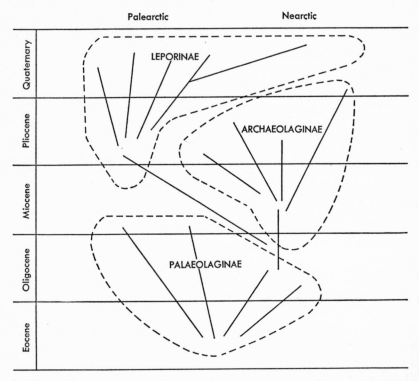

FIGURE 26. RADIATIONS CLASSIFIED AS SUBFAMILIES IN THE LEPORIDAE

The lineages shown are schematic, suggesting the radiation patterns but not particular taxa. Further explanation in text. (Schematized from more detailed representation of particular taxa by Dawson, 1958).

and the originally Palearctic radiation eventually replaced all other lineages in the New World as well as the old. There are other complications, for example, by spread into Africa and southern Asia, but on the whole the three radiations are distinguishable and can be defined on morphological evidence as subfamilies. Those subfamilies do objectively occur as units in the phylogeny, and they are "real," that is, semiarbitrary, higher categories, all three of comparable nature and scope.

As everyone knows, the Leporidae are extremely numerous in individuals. They are only moderately diversified, with nine recent genera and probably roughly comparable diversification at any one time from the Oligocene onward. Divergence within the family is slight, distinctly less than the average for mammalian families. All hares and rabbits are much alike, and have been throughout the history of the group. Divergence from the only ordinally related group, that of the pikas, Ochotonidae, is now greater than that within the Leporidae, but is less than the average for related mammalian families, and in the mid-Tertiary the two groups were so similar that they are extremely difficult to diagnose. On the usual criteria of divergence, gaps, and diversity, when compared with the arrangement of other orders, it would be fully justified to consider the Leporidae and Ochotonidae as subfamilies of one family. Since, however, they are the highest recognizable infraordinal taxa of their order and since it is convenient to rank the leporid subgroups as subfamilies, family rank is at present universally accorded to leporids and ochotonids.

The higher taxon, Lagomorpha or Duplicidentata, that includes both Leporidae and Ochotonidae was until comparatively recently classified as a suborder of Rodentia. Empirical comparison of a limited suite of characters indicated a smaller gap between lagomorphs and (other or true) rodents than is usually granted ordinal rank. Nevertheless the gap is greatly increased when other characters are taken into account, and it becomes apparent that many, if not all, of the special resemblances are convergent. With convergence excluded, the gap is evidently of ordinal magnitude. The conclusion that the Lagomorpha and the Rodentia had quite different origins and should be ranked as separate orders is strengthened by fossils, which show no indication of common ancestry at less than ordinal level, even though the ancestry has not actually been identified for either one.

The order Carnivora exemplifies definable semiarbitrary higher taxa developed by successive-replacing and contemporaneous radiations, as in the Leporidae, but at higher categorical levels and with much greater divergence and diversification. The phylogeny that now seems most probable is shown at the family level in Figure 27. Some of the classically recognized families (especially Viverridae, Procyonidae, and Mustelidae) have more than average internal divergence, and it has been proposed for them and also for some of the extinct families that they be split into a greater number of families. It is, however, fairly certain that the classical families are "good" taxa, for example, by monophyly, at some level, and I think it more convenient to continue to rank them as families. There is also a large gap, both in adaptation and in the known fossil record, between terrestrial (Fissipeda) and aquatic (Pinnipedia) carnivores, and this gap is sometimes considered of ordinal magnitude. Nevertheless the pinnipeds were almost certainly derived from early fissipeds, and I think the divergence may more conveniently continue to be ranked as subordinal.

Three radiations are evident in the pattern of Figure 27, and these are definable and now generally recognized as semiarbitrary suborders. One family lineage in the early terrestrial (creodont) radiation, that of the Miacidae, gave rise to a later terrestrial (fissiped) radiation, which eventually entirely replaced the first. The problem of subordinal assignment of the Miacidae themselves has already been discussed (Chapter 4) as an example of horizontal and vertical classification and compromise between the two. The third (pinniped) radiation occurred largely or entirely contemporaneously with the second but involved a radical change in ecology, shift from terrestrial to aquatic adaptive zones. Divergence and diversity within that radiation are much less than within the fissiped radiation.

Most of the carnivores in the basic Paleocene-Eocene radiation developed carnassial teeth and (presumed) predaceous habits, but the various lineages did so in different ways. The point is not quite certain, but the earliest Carnivora probably were not carnassial-predaceous, in fact were not truly carnivorous. The most common habitus or adaptive type of the order seems to have developed partly divergently and partly in parallel *after* the order, as a monophyletic taxon, arose. Even after it had developed, that habitus was later altogether lost in some Carnivora (for example, the panda) and profoundly modified and obscured in many others (for example, all the pinnipeds, but also a number of fissipeds). These shifting adaptive

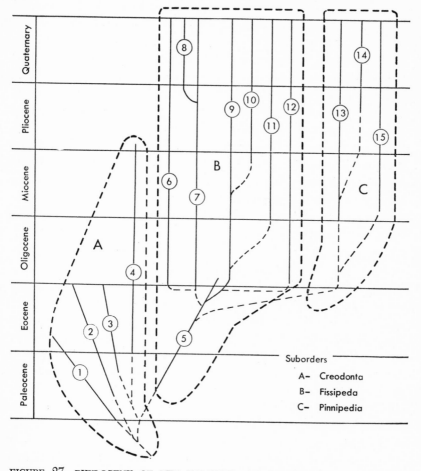

FIGURE 27. PHYLOGENY OF THE FAMILIES AND SUBORDERS OF CARNIVORA

The three suborders are enclosed in heavy broken lines. Each more or less vertical continuous line represents a family; connections in broken lines are inferred from good evidence but not represented by known fossils. The Arctocyonidae (1) are also closely allied to some other orders, and their placing in the Carnivora, although usual, is open to some question.

Families are as follows: 1, Arctocyonidae; 2, Mesonychidae; 3, Oxyaenidae; 4, Hyaenodontidae; 5, Miacidae; 6, Felidae; 7, Viverridae; 8, Hyaenidae; 9, Canidae; 10, Ursidae; 11, Procyonidae; 12, Mustelidae; 13, Otariidae; 14, Odobenidae; 15, Phocidae.

characteristics of the order are in marked contrast to the Lagomorpha, for example, in which a single adaptive type arose with the beginning of the order and was only somewhat intensified in all later members.

The Primates also show successive and contemporaneous radiations that produced semiarbitrary taxa at varying categorical levels. The pattern is highly complex in detail, and it can be taxonomically interpreted not only in terms of radiations, expansions, and contractions of the various groups but also in terms of grades, which correspond rather closely with taxa. Representation of the phylogenetic, radiation pattern in Figure 28 is highly diagrammatic. The actual lines drawn do not faithfully represent particular lineages or temporal taxa, but only symbolize the pattern impressionistically. Even as an impression, most of the complex (and not too well documented) detail is omitted within the Prosimii.

The basic radiation, which included much more divergence than any one later radiation, occurred in the Paleocene and Eocene. Numerous rapidly divergent lines soon terminated, some already in the Paleocene and most of them before the Oligocene. Few of them are known by more than jaws and teeth, but the teeth show some remarkable specializations unlike anything found in later primates. Others are more orthodox in comparison with recent primates. They include tree-shrewlike, lemurlike, and tarsierlike forms, and it was undoubtedly this complex, in a very broad sense, that gave rise to the surviving tree shrews, lemurs, lorises, and tarsiers, which have not advanced markedly over their Paleocene-Eocene forerunners. The current taxonomic solution is to include these relatively conservative forms and their early allies in the same suborder, Prosimii, and to expand that suborder basally into a horizontal taxon including also the aberrant, not even approximately ancestral early lines. Lineages among the less aberrant members of that same complex must also have given rise to the higher primates, the various members of the suborder Anthropoidea. There is, however, no present question of extending any taxon of the Anthropoidea vertically into the basal radiation complex, because no sufficiently clear connection of lineages between the two has yet been made.

The monkeys, properly speaking, evolved in two rather similar and simultaneous radiations that occurred in (partial) parallel from prob-

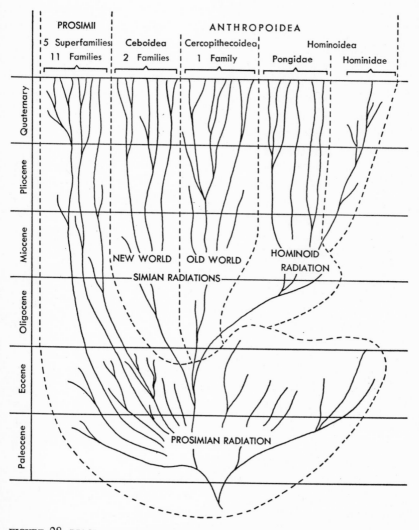

FIGURE 28. DIAGRAM OF PHYLOGENY AND CLASSIFICATION OF THE PRIMATES

The lineages shown are schematic or impressionistic of the general pattern and do not represent particular taxa. Some infraordinal taxa, at different levels as labeled above, are enclosed in broken lines.

ably similar prosimian ancestors but with wide geographic separation: one radiation in the American tropics and one in the Old World.
In large part simultaneous but probably beginning somewhat later
was another Old World radiation, that of the ape and human ancestries,
separated from the Old World monkey radiation ecologically rather
than geographically.[6] These three radiations produced comparable
higher taxa, each of which is, however, much less divergent and diverse than the prosimian radiation. The latter is classified as a suborder, while the products of the three later radiations are classified
in one suborder, Anthropoidea, and ranked as superfamilies: Ceboidea
(New World monkeys), Cercopithecoidea (Old World monkeys),
and Hominoidea (apes and men).

No clear-cut diagnostic adaptation or "heritage" distinguishes the
order Primates as a whole, or specifically the primitive primates, from
other primitive placental mammals. Later developments in intelligence, manipulation, socialization, and so on, *were* later and are
absent in the more primitive groups even among those still living.
The tree shrews, although some relationship to the Primates is generally admitted, are so primitive, so near a general ancestral placental,
that they were formerly referred to the Insectivora and still are by
a few taxonomists. Nevertheless and in spite of marked divergence
within the grade, a general prosimian grade can be recognized as
distinct from either a still more primitive ancestral placental grade
or the more advanced monkey or simian grade. The simian grade was
reached separately and in parallel by ceboids and cercopithecoids.
It is characterized by, among other things, reduction and stabilization of dental formula, and some advances in brain structure and
behavioral characteristics. The hominoid grade probably arose from
early members of the simian grade, although the exact connection
is not clear. It involved further development of the brain, loss of
the tail, increase in size, somewhat specialized omnivorous teeth, and
postural changes, among other details. The Hominidae, although forming a taxon within the higher taxon Hominoidea, deserve (we hu-

[6] In most or all such cases the initial separation was probably geographic, i.e.,
the ancestral stocks of the two radiations were probably allopatric. In this case,
however, the two were extensively sympatric through most of their radiation.
One did not replace the other, because they had different adaptive and ecological
statuses.

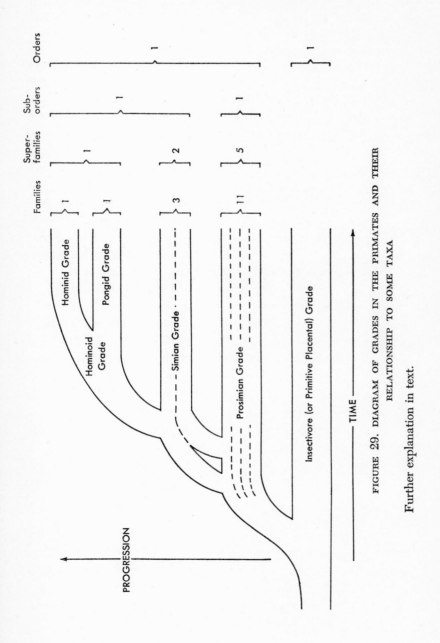

FIGURE 29. DIAGRAM OF GRADES IN THE PRIMATES AND THEIR RELATIONSHIP TO SOME TAXA

Further explanation in text.

mans are aware) separate grade recognition for their eventual development of fully upright posture, tool-making, greatest intelligence,
and speech—perhaps in that order.

Each of those grades corresponds rather closely, perhaps to the
point of identity, with a taxon. The simian grade was not monophyletic
in the strictest sense, but almost certainly was so by criteria discussed
in Chapter 4 and considered adequate for classification in one taxon.
The grades as here listed do not, however, all correspond with taxa
at the same level. On criteria of divergence, diversity, and compara-
bility the hominid grade characterizes (or even arose *within*) one
family; the hominoid grade marked the rise of a superfamily; the
simian grade is shared by two superfamilies; and the prosimian grade
is equivalent to a suborder. (It would be possible to designate lesser
grades corresponding to lower taxa within the Prosimii.) These relationships are shown diagrammatically in Figure 29.

The last example to be treated explicitly here, that of the order
Rodentia, is much the most complex, and it presents features and
problems such as frequently occur in particularly large and taxonomically difficult groups. The phylogeny shown in Figure 30 is based
essentially on the work of A. E. Wood (1955, 1958,[7] 1959), although
I have rearranged the taxa to some extent and have drawn the lines
differently. Wood runs together the lower lines in more treelike fash-
ion. My representation is not meant to imply that the 13 lines appeared
mutationally or at precisely the same time. It indicates symbolically
that they did develop rapidly and without known or evident sequence
or determined ancestral-descendant relationships among any of them.
Apart from that point, the lines shown are not impressionistic but
each represents a definite, known and named family. The names of
the individual families, 45 in number, are not necessary for present
purposes, but are given by Wood (1958, 1959) and by Simpson (1959a,
there keyed to a phylogenetic tree little different from present Figure
30). Discussion here can be in terms of what I treat as early separated basal stocks and have numbered 1–13 in Figure 30.

In spite of inevitable minor differences, mostly matters of splitting

[7] Many issues of the journal in which this paper appeared are misdated, and
"Wood, 1958" was really published too late in 1959 for consideration in Simpson,
1959a. The changes now made do not, however, affect the earlier essential conclusions.

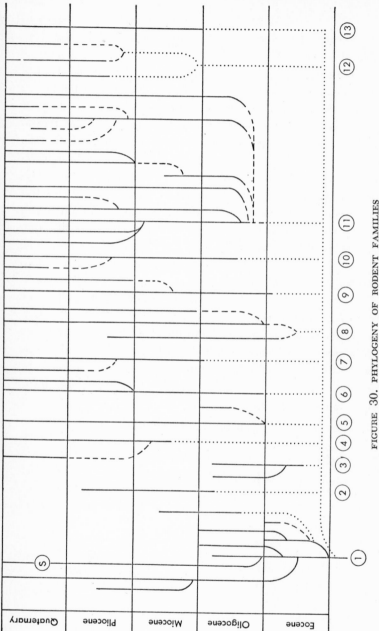

FIGURE 30. PHYLOGENY OF RODENT FAMILIES

The solid, more or less vertical lines represent the families and the temporal distribution of their known members. Phylogenetic connections in broken lines are not precisely documented by fossils but are based on strong comparative evidence. Connections in dotted lines are more inferential. The numbers at the bottom indicate the basal stocks, as discussed in the text. Their sequence is not meant to be significant except that 1 is most primitive and is believed to be near the source of all the others. (Based on data from Wood, especially 1958, 1959, but considerably modified in form of presentation.)

or lumping, the rodent families are reasonably well-defined and, in recent years, stable in classification. The order as a whole is also well-defined, certainly a valid evolutionary taxon now that the Lagomorpha have been removed, and universally accorded ordinal rank. That is a great taxonomic achievement and a highly satisfactory situation as far as it goes. It would, however, be convenient for terminology, if nothing else, to have some taxa between the one order and the 45 families, and in fact glirologists have used such intermediate taxa for well over a hundred years. The classical arrangement, formalized by Brandt (1855), recognized three suborders—Sciuromorpha, Myomorpha, and Hystricomorpha—based essentially on single character-complexes in the arrangement of the masseter muscle. Although still more or less current,[8] that arrangement is a typological classification of characters and not an evolutionary classification of animals. The animals plainly show that each of the three arrangements arose, with significant variations, independently in various different ancestral groups from a more primitive arrangement unknown to Brandt. They certainly do not warrant recognition of three, and only three, subordinal taxa. Yet no fully acceptable substitute has been proposed.

The usual alternative has been to multiply the number of suborders, until Wood (1958) has 11 in an arrangement that he shows to be logical but does not himself definitely accept. In terms of the numbered basal lineages of Figure 30, that arrangement is as follows:

Protrogomorpha – Part of 1 (all but the line marked "s").
Caviomorpha – 11.
Myomorpha – 6, 7, 8, 9.
Perhaps new suborder, perhaps Myomorpha – 10.

———

Sciuromorpha – Part of 1 (the line marked "s").
Castorimorpha – 5.
Theridomorpha – 3.
New suborder (not named) – 2.
New suborder (not named) – 4.
Hystricomorpha – 12.
Bathyergomorpha – 13.

[8] I still used it reluctantly in 1945, although with strongly stated reservations. I now think even such grudging and conditional acceptance entirely unjustified.

Alternatively, Wood suggests possible preservation of only the sub-orders above the line and simply listing the other families (some under and some not under superfamilies) as "Groups not allocated to suborders." He points out advantages and disadvantages in both arrangements, but does not say which he prefers. He then concludes:

In summary, it is suggested that the present situation, where there are a considerable number of families of rodents that seem to be taxonomically isolated, may be an actual expression of a basic fact in rodent evolution, and that these small groups may never have had a major importance, and have common ancestors only within the primitive suborder, the Protrogo-morpha [bottom of line 1 in Figure 30].

I agree that that is the situation: in other words, that the phylo-genetic pattern of this order simply cannot be divided in an accept-able way into a small number of similar or comparable suborders. Reference again to Figure 30 shows that 11 of the 13 basal lines that I recognize diversified little or not at all at the family level. (Two remain single families; seven split into two families; and two become three families.) Yet there is little good, and no conclusive, evidence for considering any two of those basal lines especially related *inter se*.[9] These, at least, are comparable taxa at some level, but with re-gard to these lines, alone, the appropriate level is evidently not higher than superfamily. Lines 1 and 11, however, are radically different. The nine families descended from 1 evolve in a pattern of simple branch-ing, expansion, and then restriction as most of the families become extinct. Line 11 suddenly splits basally into five or six lines, very much as the whole order split into 13 at an earlier date, and the group even-tually expands to 14 families.

The phylogeny developed from line 1 and that from line 11 are not very closely comparable, but they are both such as could be con-sidered suborders if the rest of the order were ignored. It is, more-over, highly desirable or even a practical necessity to have names by which to refer to these groups of certainly related families. They do, in fact, have names given by Wood: Protrogomorpha (plus the Sciuridae, which Wood removes for reasons I find inadequate) for the group from line 1, and Caviomorpha for the group from line 11.

[9] Wood does still consider my lines 6, 7, 8, and 9 especially related, but the evidence is exiguous and he is perhaps only leaning over backwards in trying to avoid splitting of a classical suborder.

It is contrary to usage and to the whole logic of the hierarchy to call groups 1 and 11 suborders and to call 2–10 and 12–13 superfamilies without allocation to suborders. Yet the two kinds of groups are not comparable, and calling all either suborders or superfamilies is unsatisfactory for that reason. Nature here simply has not presented us with comparable taxa at any level between family and order, and if such taxa are used the only solution is to admit into the same category taxa that are not comparable. Then 1–13 of Figure 30 would all be suborders (with superfamilies in 1 and 11, but not in the others) or all superfamilies (none of them aggregated into suborders).

As for the evolutionary factors involved, the order arose with one basic adaptation, for gnawing, so effective that Rodentia quickly replaced other gnawers and rapidly diversified into many groups over Africa, Eurasia, and North America. All descendants retained the gnawing adaptation. Almost all tended to perfect it progressively and in parallel in several different ways, notably in the masseter arrangement. For the masseter, three alternative specializations, corresponding with Brandt's typological groups, were possible and apparently equally effective. One or the other appeared, seemingly almost at random, in various early lines (with exceptions only in group 1). Once one or the other of the specialized arrangements had evolved, it was generally retained in all descendants of that particular line, although sometimes with considerable further modification. There were also widespread, although considerably less general, parallel specializations in other respects, notably in the development of hypsodont and crested cheek teeth. In other respects the rodents were remarkably divergent and eventually developed a variety of adaptive types practically unparalleled within a single order. That divergence, occurring in many different large areas, also involved repeated and sometimes intricate convergence between rodents of similar ecology but in different regions.

Rodents reached South America belatedly, around the end of the Eocene. They there encountered only limited and mostly ineffective competition among some marsupials and small ungulates, which soon either became extinct or specialized in ways less competitive with rodents. The rodents thus had an almost clear field in a region highly varied ecologically and of continental size. They diverged rapidly and then diversified markedly, developing the special phylogenetic pattern from line 11 of Figure 30.

It has been emphasized throughout that in evolutionary classification the taxa are either made consistent with a reconstruction of phylogeny or with phylogenetic processes likely to have produced an observed result—or both, for those approaches are closely and often inseparably related. In any case the taxa reflect evolutionary factors involved in their origin and development, and this highly meaningful relationship is prominent among the several reasons for insisting, as far as practicable, on the evolutionary approach to classification. By way of underlining that relationship and of relating taxonomy to the whole field of systematics and evolution, something needs finally to be said about the evolutionary meaning of taxa or their natures in evolutionary terms. It is impractical here to go into most of the basic or elemental processes, for instance those at the genetic level, the various isolating mechanisms, or the factors of natural selection. What can usefully be done in concluding this extended essay is to relate generalizations about taxa to generalizations about the histories of populations of whole organisms, for the most part in terms of adaptation, which is the most significant evolutionary factor in this connection.

Evolutionary species have been defined (Chapter 5) directly by the separate and unitary nature of their evolutionary roles. Equivalence, at any one place, with minimal separable ecological niches defines the roles and their adaptive nature. Infraspecific taxonomic units, in a great majority of instances allopatric within the distribution of the whole species, usually represent still closer adaptation to microenvironments, modified by such factors as gene flow, genetic drift, or mutation pressure. The adaptive status of these infraspecific populations, or of subspecific taxa, is not clearly differentiated from that of adjacent populations or of the species as a whole, and such divergence as occurs is unlikely to be sustained for long spans of time unless, in fact, such an infraspecific unit evolves into a species. Essential duplication of adaptations by different species is unusual and temporary unless the species are fully allopatric and separated by a geographical or other ecological barrier.

Taxa of higher categorical rank involve one or, as a rule with few exceptions, both of two adaptive sequences. First, from a common ancestry, either unispecific or in a group of species still capable of gene exchange,[10] multiple specific lineages with different adaptive roles evolve. That difference in roles may, in principle, be only distributional and hence external to the organisms themselves, but in actuality it almost always also involves somatic differences in the organisms. Second, any one lineage may diverge adaptively so far from any others as to constitute a higher taxon in itself. Such a taxon is still polytypic in the sense that it includes more than one *successional* species, each adaptively more or less distinct from the others.[11] It is clear that a combination of these two processes, of diversification and divergence, is the rule.

Most higher taxa involve some basic adaptation that evolves either coevally with the taxon itself, that is, at the base of what *later* becomes a higher taxon, or with more or less parallelism among its early lineages. The basic adaptation may involve essentially a single structure or limited set of functionally closely related structures, like the gnawing adaptation of rodents or the flying adaptation of bats. At the other extreme, it may involve a whole complex of many different characters and structures, like the aquatic adaptation of whales. That is obviously not a dichotomy, but a continuous range from single-character through all degrees of multiple-character basic adaptation. Such origins of higher taxa by basic adaptation usually occur when there is a shift from one fairly distinct adaptive zone to another.

The eventual rank of the taxon thus initiated is usually proportional

[10] Monophyly in this strictest sense undoubtedly is basic to the evolutionary process here in question. As has been sufficiently noted, it is not always a practical criterion for defining the taxa of formal classification, which is a different although related question.

[11] Hence the monotypic taxa of classifications are such only because of incomplete knowledge. In principle they include multiple empty sets, taxa that occurred in nature but have not been found and classified. The only possible exceptions would result from some one-step process that forthwith produced a gap of generic or higher magnitude, if, indeed, that ever really occurs.

to the degree of distinction of the zone entered, hence the amount of basic divergence involved, and to its scope or number of subzones and niches, hence the opportunities for diversification within it. Among the vertebrates the shift from the aquatic to the terrestrial zones involved great basic divergence and also entrance into a very broad, extremely varied zone. Several classes, dozens of orders, and many thousands of species eventuated. In the rodents, bats, and whales, given as examples in the previous paragraph, an order eventuated in each case. The rank eventually achieved is not, however, a necessary function of the kind of adaptation involved. For example, gnawing adaptation similar to that of the rodents has occurred within several other orders (for example, marsupials, primates, various ungulates) but with less initial divergence and subsequent diversification so that in those groups it gave rise only to taxa around the rank of family or even genus.

Adaptations involved in the origin of higher taxa differ both in breadth, that is, the number of different accessible ecological situations in which the adaptation is advantageous (or pertinent), and in retention, that is, the extent to which the adaptation continues to characterize the whole resulting higher taxon. In general, and rather obviously, the broader the adaptation the more likely it is to be retained, but this is not a perfect correlation. The gnawing adaptation of rodents, flying adaptation of bats, and aquatic adaptation of whales were broad and were retained without exception in all members of the resulting higher taxa. On the other hand, the terrestrial locomotor adaptation of early amphibians was certainly one of the broadest possible, and yet it was lost by many of their descendants. Those descendants did not, however, lose some other parts of the multiple terrestrial adaptation, for example, air-breathing, even when they ceased actually to be terrestrial. Retention of basic adaptations is the main source of characters in common. Consideration from this point of view makes clearer why characters in common are generally useful in recognizing evolutionary taxa, but do also rather frequently prove inadequate in that role.

Breadth and retention of adaptation form a continuous scale which can be related to categories and put in those terms. Thus there are species-specific, genus-specific, family-specific, and so on even to kingdom-specific adaptations and corresponding somatic characteristics, although for various reasons these become somewhat more liable to exceptions as the taxonomic scale is ascended.

In such higher taxa as arise by a fairly clear-cut and single adaptation, there is commonly a slower and subsequent coadaptation of other characteristics in relation to the basic adaptation. Coadaptation may occur differently in different lineages, but it frequently involves extensive parallelism. Many of the more or less parallel trends among rodents, such as each of the three masseter arrangements, hypsodonty, and lophiodonty, may be considered coadaptive with respect to the basic gnawing adaptation.

On the other hand, when multiple adaptations, or adaptive systems, are involved in the rise of a higher taxon, those basic adaptations are themselves likely to evolve in parallel in lineages around the origin of the group. Different parts of the complex commonly evolve at different times and rates within one lineage, giving the effect that de Beer (1954) calls "mosaic evolution." Thus in *Archaeopteryx*, de Beer's example, some characteristics were still fully reptilian, some already fully avian, and some intermediate. Nevertheless the pieces in the "mosaic" are not inserted or shuffled independently to produce the new picture. They, too, coadapt even if only in a loose sense, and the whole organism remains integrated. The same parts of the complex may also evolve at different times and rates in different lineages. There is then no one or limited time within which the whole adaptive complex of the nascent higher taxon appears even in one lineage and still less in a group of lineages with parallel evolution of the complex. This situation is strikingly exemplified in the origin of the class Mammalia, the total adaptive complex, facies, or grade of which was in the process of parallel evolution in various lineages during the whole tremendous span from the Permian into the Cretaceous (Simpson, 1959b, 1960b). That gives rise to a problem of classification that can only be solved by arbitrary selection for purposes of diagnosis of some one character within the complex, even a comparatively slight one if it is relevant to the basic adaptation and is convenient for practical application.

At the same time that parallel trends are going on throughout a higher taxon, either in its basic adaptations or in later coadaptations, divergence occurs among included taxa at each lower level down to the species. Thus while the eventually class-specific adaptations of the Mammalia were arising, there were also order-specific, family-specific, genus-specific, and species-specific adaptations in the included taxa.

The slow attainment of a complex basic adaptation grades into and is not clearly distinguishable from instances in which one can hardly say that there is a diagnostic or basic adaptation in the origins of the taxon but only progressive trends, or improvements in Huxley's (for example, 1957) sense within the taxon during much or all of its history. Both these intergrading processes are particularly likely to produce grades, generally polyphyletic at low taxonomic levels but monophyletic near the level of the taxon actually arising. The general nature of these phenomena and their relationship to taxa are exemplified by the primates, as discussed in the preceding section. In that example all the important grades have survived and there was little replacement of a lower or older grade by one higher or later. That is because this was not, in fact, a simple matter of improvement of adaptation within the same adaptive zone but also and simultaneously of development of different adaptive zones, with little overlap or competition. In the example of the Carnivora, however, the creodonts and fissipeds, which also can be considered grades although of a somewhat different kind, there was enough overlap of zones to permit competition and final replacement of the lower grade, which is also a taxon, by the higher, another taxon of the same rank.

There is a balance in higher taxa between progression of the group as a whole or its divergence from others at the same level, and its diversification with divergence among its included lesser taxa. The Lagomorpha show some over-all progression, but little internal divergence. The Carnivora have little over-all progression but great internal divergence. The Rodentia exhibit both phenomena in marked degree, and almost all higher taxa show each to some degree. A characteristic form of diversification is by adaptive radiation, which frequently produces groups that are only semiarbitrary and that are closely comparable so that they can be satisfactorily defined as taxa at the same level. A number of examples have been given, such as the subfamilies of Leporidae and suborders of Carnivora. In some instances, however, a single, comparatively rapid basal adaptive radiation produces groups strongly different in the balance of progression and diversification, hence not well comparable and difficult or impossible to place satisfactorily as taxa at any one level in the hierarchy. The rodent groups between family and order are a striking example.

When two or more higher taxa near or at the same categorical level arise from a common ancestry, they generally do so allopatrically (in different regions, like Ceboidea and Cercopithecoidea), allochronously (at different times, like Prosimii and Anthropoidea), or both (like Ceboidea and Hominoidea; other examples of all these relationships also occur in the preceding section). In any case it often happens that the expanding groups, after they are established, become both sympatric and synchronous. If, in the meantime, they have become adaptively quite distinct, nothing but a general enrichment of the fauna ensues. Every fauna does include numerous different taxa of ultimate common ancestry. If, however, there is still enough adaptive similarity to involve competition, there is an interplay of two processes: finer subdivision of adaptive zones with their parceling out among the various groups, and replacement, or extinction and relay, of one group by another. Both processes occurred on a wide scale, for instance, when northern rodents expanded to South America late in the Cenozoic and came into competition with some of the older native rodents there.

Convergence is most likely to occur and is most intensive between taxa of allopatric radiations and while there is little or no expansion of taxa from one radiation into the area occupied by the other. It is improbable that any really close convergence has evolved while the taxa in question were sympatric, and it is at least unusual for closely convergent taxa ever to become sympatric and remain so for any considerable length of time. That is one of the reasons why zoogeography is especially pertinent in evolutionary classification, which demands sharp discrimination between convergence and homology. The classic examples of large-scale multiple convergence are the old native mammalian faunas of Australia and South America. Each continent was the scene of an extensive, complex adaptive radiation, largely isolated from radiations on the other continents. Both in Australia and in South America numerous groups developed with strong convergence toward African-Eurasian-North American mammals. When for one reason or another convergent vicars of different geographic origin did become sympatric, the convergent Australian and South American taxa soon became extinct, for example, thylacines when true dogs came into Australia, or native South American ungulates when perissodactyls and artiodactyls spread to that continent.

It is with the results of all these processes—and more!—that taxonomy must deal. The interplay of the processes is so complex that it produces a bewildering array of phylogenetic patterns among all animals, of associations of contiguity and of similarity. Evolution has not provided a series of nicely nested boxes into which all its products fit. Any taxonomic approach that presupposes a simple, natural arrangement that can be found out by either strictly a priori or strictly empirical means is doomed to failure by the nature of the materials to be classified. Yet evolution has not produced chaos or a hopeless tangle, either. It is a completely orderly process, but ordered by numerous interacting, balancing, at times even conflicting principles. The principles can be learned, if only by successive and sometimes groping approximations. There is a science of taxonomy, which adapts the principles of evolution to its special field and adds principles of its own. The principles of taxonomy are applied in classification, which is also an art with canons of taste, of moderation, and of usefulness.

Works Cited

Adanson, M. 1763. Familles des plantes. Paris, Vincent.

Allee, W. C., A. E. Emerson, O. Park, T. Park, and K. P. Schmidt. 1949. Principles of animal ecology. Philadelphia, Saunders.

Allen, J. A. 1919. Notes on the synonymy and nomenclature of the smaller spotted cats of tropical America. Bull. Amer. Mus. Nat. Hist., 41: 341–419.

Amadon, D. 1949. The seventy-five per cent rule for subspecies. Condor, 51: 250–258.

Ashton, E. H., M. J. R. Healy, and S. Lipton. 1957. The descriptive use of discriminant functions in physical anthropology. Proc. Roy. Soc., B, 146: 552–572.

Bader, T. S. 1958. Similarity and recency of common ancestry. Systematic Zoology, 7: 184–187.

Baer, K. E. von. 1828. Ueber Entwicklungsgeschichte der Thiere. Beobachtungen und Reflexion. Königsberg, Wilhelm Koch.

Bather, F. A. 1927. Biological classification past and future. Quart. Journ. Geol. Soc., 83: lxii–civ.

Beckner, M. 1959. The biological way of thought. New York, Columbia Univ. Press.

Belon, P. 1555. L'histoire de la nature des oyseaux. Paris, Guillaume Cavellat.

Bigelow, R. S. 1958. Classification and phylogeny. Systematic Zoology, 7: 49–59.

Blackwelder, R. E. 1959. The present status of systematic zoology. Systematic Zoology, 8: 69–75.

Blackwelder, R. E., and A. Boyden. 1952. The nature of systematics. Systematic Zoology, 1: 26–33.

Blainville, H. M. D. de. 1816. Prodrome d'une nouvelle distribution systématique du règne animal. Bull. Sci. Soc. Philom. Paris, sér. 3, 3: 105–124.

Block, K. 1956. Zur Theorie der naturwissenschaftlichen Systematik unter besonderer Berücksichtigung der Biologie. Bibl. Biotheor., 7: 1–138.

Bonnet, C. 1745. Traité d'insectologie. Paris, Durand.

Boyden, A. 1947. Homology and analogy. Amer. Midland Naturalist, 37: 648–669.

—— 1953. Fifty years of systematic serology. Systematic Zoology, 2: 19–30.

—— 1958. Comparative serology: aims, methods, and results. Serological and Biochemical Comparison of Proteins, XIV Annual Protein Conference: 3–24.

Brandt, J. F. 1855. Beiträge zur näheren Kenntniss der Säugethiere Russlands. Mém. Acad. Imp. Sci. St.-Pétersbourg, ser. 6, 9: 1–375.

Brinkmann, R. 1929. Statistisch-biostratigraphische Untersuchungen an mitteljurassischen Ammoniten über Artbegriff und Stammesentwicklung. Abh. Ges. Wiss. Göttingen, math.-phys. Kl., N.F., 8: part 3.

Brisson, M. J. 1762. Regnum animale in classes IX. 2nd. ed. Paris, Cl. Joannem-Baptistam Bauche.

Bulman, O. M. B. 1955. Graptolithina. Treatise on Invertebrate Paleontology, Part V.

Burma, B. J. 1948. Studies in quantitative paleontology: I. Some aspects of the theory and practice. Jour. Paleont., 22: 725–761.

Cain, A. J. 1953. Geography, ecology and coexistence in relation to the biological definition of the species. Evolution, 7: 76–83.

—— 1954. Animal species and their evolution. London, Hutchinson.

—— 1955. Not the superspecies. Systematic Zoology, 4: 143–144, 159.

—— 1958a. Logic and memory in Linnaeus's system of taxonomy. Proc. Linnean Soc. London, 169 session: 144–163.

—— 1958b. Chromosomes and taxonomic importance. Proc. Linnean Soc. London, 169 session: 125–128.

—— 1959a. The post-Linnaean development of taxonomy. Proc. Linnean Soc. London, 170: 234–244.

—— 1959b. Deductive and inductive methods in post-Linnaean taxonomy. Proc. Linnean Soc. London, 170: 185–217.

—— 1959c. Taxonomic concepts. Ibis, 101: 302–318.

Cain, A. J., and G. A. Harrison. 1958. An analysis of the taxonomist's judgement of affinity. Proc. Zool. Soc. London, 131: 85–98.

Calman, W. T. 1940. A museum zoologist's view of taxonomy. In Huxley, 1940: 455–459.

Christensen, B., and C. O. Nielsen. 1955. Studies on Enchytraeidae, 4. Preliminary report on chromosome numbers of seven Danish genera. Chromosoma, 7: 460–468.

Clark, W. E. Le Gros. 1959. The antecedents of man. Chicago, Quadrangle Books.

Cloud, P. E., Jr. 1959. Paleoecology—retrospect and prospect. Jour. Paleont., 33: 926–962.

Colbert, E. H. 1941. A study of Orycteropus gaudryi from the Island of Samos. Bull. Amer. Mus. Nat. Hist., 78: 305–351.

Cope, E. D. 1875. On the supposed *Carnivora* of the Eocene of the Rocky Mountains. Proc. Acad. Nat. Sci. Philadelphia, 27: 444–448.

Cuvier, G. 1835. Le règne animal distribué après son organization. 2nd ed. Paris, Crochard et Cie.

Darlington, P. J. 1957. Zoogeography: the geographical distribution of animals. New York, Wiley.

Darwin, C. 1859. On the origin of species by means of natural selection or the preservation of favoured races in the struggle for life. London, Murray. (For all subsequent revisions see the variorum edition: Morse Peckham, ed., 1959. Philadelphia, Univ. of Pennsylvania Press.)

Dawson, M. R. 1958. Late Tertiary Leporidae of North America. Univ. Kansas Paleo. Cont., Vertebrata, art. 6: 1–75.

De Beer, G. R. 1951. Embryos and ancestors. Rev. ed. Oxford, Oxford Univ. Press.

—— 1954. *Archaeopteryx lithographica*. A study based upon the British Museum specimen. London, British Mus. (Nat. Hist.).

Delacour, J., and E. Mayr. 1945. The family Anatidae. Wilson Bull., 57: 3–55.

Dennler, J. G. 1939. Los nombres indígenas en guaraní de los mamíferos de la Argentina y países limítrofes y su importancia para la sistemática. Physis, 16: 225–244.

Dethier, V. G. 1954. Evolution of feeding preferences in phytophagous insects. Evolution, 8: 32–54.

Dobzhansky, Th. 1937, 1941, 1951. Genetics and the origin of species. 1st ed. (1937). 2nd ed. (1941). 3rd ed. (1951). New York, Columbia Univ. Press.

Eagar, R. M. C. 1956. Naming Carboniferous non-marine lamellibranchs. *In* Sylvester-Bradley, 1956: 111–116.

Edwards, J. G. 1954. A new approach to infraspecific categories. Systematic Zoology, 3: 1–20.

Garstang, W. 1922. The theory of recapitulation. Jour. Linnean Soc. London, Zool., 35: 81–101.

George, T. Neville. 1956. Biospecies, chronospecies, and morphospecies. *In* Sylvester-Bradley, 1956: 123–137.

Gilmour, G. S. L. 1940. Taxonomy and philosophy. *In* Huxley, 1940: 461–474.

—— 1951. The development of taxonomic theory since 1851. Nature, 168: 400–402.

Goldschmidt, R. 1940. The material basis of evolution. New Haven, Yale Univ. Press.

Grabert, B. 1959. Phylogenetische Untersuchungen an *Gaudryina* und *Spiroplectinata* (Foram.) besonders aus dem nordwestdeutschen Apt und Alb. Abh. Senckenberg. Naturforsch. Ges., No. 498: 1–71.

Gregg, J. R. 1954. The language of taxonomy. New York, Columbia Univ. Press.

Gregory, W. K. 1910. The orders of mammals. Bull. Amer. Mus. Nat. Hist., 27: 1–524.

—— 1951. Evolution emerging. 2 vols. New York, Macmillan.

Haas, O., and G. G. Simpson. 1946. Analysis of some phylogenetic terms with attempts at redefinition. Proc. Amer. Phil. Soc., 90: 319–349.

Haeckel, E. 1875. Ziel und Wege der heutigen Entwicklungsgeschichte. Jena, Hermann Dufft.

Haslewood, G. A. D. 1959a. Species comparison as an aid in the study of the process sterols → bile salts. Ciba Symp. on Biosynthesis of Terpines and Sterols: 206–213.

—— 1959b. Comparative studies of "bile salts." 12. Application to a problem of rodent classification: bile salts of the cutting-grass, *Thryonomys swinderianus*. Biochem. Jour., 73: 142–144.

Heberer, G., ed. 1943, 1954–1959. Die Evolution der Organism. 1st ed. (1943). 2nd ed. (1954–1959). Stuttgart, Gustav Fischer.

Hedgpeth, J. W. (ed. vol. 1), and H. S. Ladd (ed. vol. 2). 1957. Treatise on marine ecology and paleoecology. 2 vols. Geol. Soc. Amer., Mem. 67.

Hennig, W. 1950. Grundzüge einer Theorie der phylogenetischen Systematik. Berlin, Deutscher Zentralverlag.

Huxley, J. S. 1938. Clines: an auxiliary taxonomic principle. Nature, 142: 219.

—— 1939. [No title; discussion of clines.] Proc. Linnean Soc. London, 1938–39: 105–114.

—— ed. 1940. The new systematics. Oxford, Oxford Univ. Press.

—— 1942. Evolution, the modern synthesis. New York, Harper.

—— 1957. The three types of evolutionary process. Nature, 180: 454–455.

—— 1958. Evolutionary processes and taxonomy with special reference to grades. Uppsala Univ. Arssks., 1958: 21–38.

Huxley, T. H. 1880. On the application of the laws of evolution to the arrangement of the Vertebrata and more particularly of the Mammalia. Proc. Zool. Soc. London, 1880: 649–662.

Imbrie, J. 1957. The species problem with fossil animals. *In* Mayr, 1957: 125–153.

Inger, R. F. 1958. Comments on the definition of genera. Evolution, 12: 370–384.

Jepsen, G. L. 1949. Selection, "orthogenesis," and the fossil record. Proc. Amer. Phil. Soc., 93: 479–500.

Johnson, M. L., and M. J. Wicks. 1959. Serum protein electrophoresis in mammals—taxonomic implications. Systematic Zoology, 8: 88–95.

Jolicoeur, P. 1959. Multivariate geographical variation in the wolf, *Canis lupus* L. Evolution, 13: 283–299.

Joseph, H. W. B. 1916. An introduction to logic. 2nd ed. Oxford, Oxford Univ. Press.

Kaplan, N. O., M. M. Ciotti, M. Hamolosky, and R. E. Bieber. 1960. Molecular heterogeneity and evolution of enzymes. Science, 131: 392–397.

Kinsey, A. C. 1936. The origin of higher categories in *Cynips*. Indiana Univ. Pub., Sci. Ser., No. 4: 1–334.

Kiriakoff, S. G. 1959. Phylogenetic systematics versus typology. Systematic Zoology, 8: 117–118.

Klein, J. T. 1751. Quadrupedum dispositio brevisque historia naturalis. Lipsiae, Jonam Schmidt.

Kleinschmidt, O. 1926. Die Formenkreislehre. Halle-S., Gebauer-Swetschke.

Krutzsch, P. H. 1954. North American jumping mice (genus *Zapus*). Univ. Kansas Pub., Mus. Nat. Hist., 7: 349–472.

Lamarck, J.-B. M. de. 1809. Philosophie zoologique. Paris, Dentu.

Lerner, I. M. 1954. Genetic homeostasis. New York, Wiley.

Linnaeus, C. 1758. Systema naturae. 10th ed. Stockholm, Laurentii Salvii.

Lull, R. S. 1953. Triassic life of the Connecticut Valley. Connecticut State Geol. and Nat. Hist. Surv., Bull. No. 81.

Marsh, F. L. 1944. Evolution, creation and science. Washington, D.C., Review and Herald Publishing Association.

Maslin, T. P. 1952. Morphological criteria of phylogenetic relationships. Systematic Zoology, 1: 49–70.

Matthew, W. D. 1909. The Carnivora and Insectivora of the Bridger basin, middle Eocene. Mem. Amer. Mus. Nat. Hist., 9: 289–567.

Matthey, R. 1952. Chromosomes de Muridae (Microtinae et Cricetinae). Chromosoma, 5: 113–138.

Mayr, E. 1931. Notes on *Halcyon chloris* and some of its subspecies. Amer. Mus. Novitates, No. 469: 1–10.

—— 1940. Speciation phenomena in birds. Amer. Naturalist, 74: 249–278.

—— 1942. Systematics and the origin of species from the viewpoint of a zoologist. New York, Columbia Univ. Press.

—— 1948. The bearing of the new systematics on genetical problems. The nature of species. Advances in Genetics, 2: 205–237.

—— 1954. Notes on nomenclature and classification. Systematic Zoology, 3: 86–89.

——, ed. 1957. The species problem. Amer. Assoc. Adv. Sci., Pub. No. 50.

—— 1958. Behavior and systematics. *In* Roe and Simpson, 1958: 340–362.

Mayr, E., E. G. Linsley, and R. L. Usinger. 1953. Methods and principles of systematic zoology. New York, McGraw-Hill.

Meglitsch, P. A. 1954. On the nature of the species. Systematic Zoology, 3: 49–65.

Merriam, C. H. 1918. Review of the grizzly and big brown bears of North

America (genus *Ursus*) with description of a new genus, *Vetularctos*. N. Amer. Fauna, 41: 1–36.

Michener, C. D. 1949. Parallelisms in the evolution of the Saturnid moths. Evolution, 3: 129–141.

—— 1957. Some bases for higher categories in classification. Systematic Zoology, 6: 160–173.

Michener, C. D., and R. S. Sokal. 1957. A quantitative approach to a problem in classification. Evolution, 11: 130–162.

Moore, R. C. 1957. Digest of proposed addition to the "Regles" of provisions recognizing and regulating the nomenclature of "Parataxa." Jour. Paleont., 31: 1181–1183.

Mossman, H. W. 1953. The genital system and fetal membranes as criteria for mammaliam phylogeny and taxonomy. Jour. Mam., 34: 289–298.

Newell, N. D. 1947. Infraspecific categories in invertebrate paleontology. Evolution, 1: 163–171.

—— 1949a. Phyletic size increase—an important trend illustrated by fossil invertebrates. Evolution, 3: 103–124.

—— 1949b. Types and hypodigms. Amer. Jour. Sci., 247: 134–142.

—— 1956. Fossil populations. *In* Sylvester-Bradley, 1956: 63–82.

Nicol, D., G. A. Desborough, and J. R. Solliday. 1959. Paleontologic record of the primary differentiation in some major invertebrate groups. Jour. Washington Acad. Sci., 49: 351–366.

Olson, E. C., and R. L. Miller. 1958. Morphological integration. Chicago, Univ. of Chicago Press.

Orton, G. L. 1953. The systematics of vertebrate larvae. Systematic Zoology, 2: 63–75.

—— 1955. The role of ontogeny in systematics and evolution. Evolution, 9: 75–83.

Osborn, H. F. 1936, 1942. Proboscidea. 2 vols. New York, Amer. Mus. Nat. Hist.

Owen, R. 1848. Report on the archetype and homologies of the vertebrate skeleton. Rep. 16th Meeting British Assoc. Adv. Sci.: 169–340.

—— 1866. On the anatomy of vertebrates. London, Longmans, Green.

Patterson, J. T., and W. S. Stone. 1952. Evolution in the genus *Drosophila*. New York, Macmillan.

Pennant, T. 1781. History of quadrupeds. London, B. White.

Pirie, N. W. 1952. Concepts out of context: the pied pipers of science. Brit. Jour. Phil. Sci., 2: 269-280.

Pontecorvo, G. 1959. Present trends in genetic analysis. New York, Columbia Univ. Press.

Quinn, J. H. 1955. Miocene Equidae of the Texas Gulf Coastal Plain. Univ. of Texas Pub. No. 5516: 1–102.

Remane, A. 1956. Die Grundlagen des natürlichen Systems, der vergleichenden Anatomie und der Phylogenetik. Leipzig, Geest and Portig.

Rensch, B. 1929. Das Prinzip geographischer Rassenkreise und das Problem der Artbildung. Berlin, Gebrüder Borntraeger.

—— 1954. Neuere Probleme der Abstammungslehre. Die transspezifische Evolution. 2e Aufl. Stuttgart, Ferdinand Enke.

—— 1960. The laws of evolution. In The evolution of life, S. Tax, ed., vol. I of "Evolution after Darwin." Chicago, Univ. of Chicago Press, pp. 95–116.

Rhodes, F. H. T. 1956. The time factor in taxonomy. In Sylvester-Bradley, 1956: 33–52.

Roe, A. 1946. Artists and their work. Jour. Personality, 15: 1–40.

—— 1953. The making of a scientist. New York, Dodd, Mead.

Roe, A., and G. G. Simpson, eds. 1958. Behavior and evolution. New Haven, Yale Univ. Press.

Rosa, D. 1931. L'Ologénèse; nouvelle théorie de l'évolution et de la distribution géographique des êtres vivants. Paris, Alcan.

Scheffer, V. B. 1958. Seals, sea lions, and walruses. Stanford, Stanford Univ. Press.

Schindewolf, O. H. 1950. Grundfragen der Paläontologie. Stuttgart, E. Schweizerbart'sche Verlagsbuchhandlung (Erwin Nägele).

Schmalhausen, I. I. 1949. Factors of evolution; the theory of stabilizing selection. Trans. by I. Dordick. Ed. by Th. Dobzhansky. Philadelphia, Blakiston.

Scott, A. J., and C. Collinson. 1959. Intraspecific variability in conodonts: Palmatolepis glabra Ulrich and Bassler. Jour. Paleont., 33: 550–565.

Shenefelt, R. D. 1959. Taxonomic "descriptions." Science, 130: 331.

Sibley, C. G. 1954. The contribution of avian taxonomy [to a symposium on subspecies and clines]. Systematic Zoology, 3: 105–110, 125.

Simpson, G. G. 1937. Super-specific variation in nature and in classification from the view-point of paleontology. Amer. Naturalist, 71: 236–267.

—— 1940. Types in modern taxonomy. Amer. Jour. Sci., 238: 413–431.

—— 1941. The affinities of the Borhyaenidae. Amer. Mus. Novitates, No. 1118: 1–6.

—— 1943. Criteria for genera, species, and subspecies in zoology and paleozoology. Ann. N.Y. Acad. Sci., 44: 145–178.

—— 1944. Tempo and mode in evolution. New York, Columbia Univ. Press.

—— 1945. The principles of classification and a classification of mammals. Bull. Amer. Mus. Nat. Hist., 85: i–xvi, 1–350.

—— 1951. The species concept. Evolution, 5: 285–298.

—— 1953. The major features of evolution. New York, Columbia Univ. Press.

—— 1954. Tendances actuelles de la systématique des mammifères. Mammalia, 18: 337–357.

—— 1959a. The nature and origin of supraspecific taxa. Cold Spring Harbor Symposia on Quantitative Biology, 24: 255–271.

—— 1959b. Mesozoic mammals and the polyphyletic origin of mammals. Evolution, 13: 405–414.

—— 1959c. Anatomy and morphology: classification and evolution: 1859 and 1959. Proc. Amer. Phil. Soc., 103: 286–306.

—— 1960a. The history of life. In The evolution of life, S. Tax, ed., vol. I of "Evolution after Darwin." Chicago, Univ. of Chicago Press. pp. 117–180.

—— 1960b. Diagnosis of the classes Reptilia and Mammalia. Evolution, [in press].

Simpson, G. G., A. Roe, and R. C. Lewontin. 1960. Quantitative zoology. Revised ed. New York, Harcourt, Brace.

Smith, H. M., and F. N. White. 1956. A case for the trinomen. Systematic Zoology, 5: 183–190.

Sokal, R. R., and C. D. Michener. 1958. A statistical method for evaluating systematic relationships. Univ. Kansas Sci. Bull., 38: 1409–1438.

Sonneborn, T. M. 1957. Breeding systems, reproductive methods, and species problems in Protozoa. In Mayr, 1957: 155–324.

Sprague, T. A., A. T. Hopwood, A. J. Wilmott, H. K. Airey Shaw, and J. Smart. 1950. Lectures on the development of taxonomy delivered in the rooms of the Linnean Society during the session 1948–1949. Linnean Soc., London. [A pamphlet of 83 pages, without name of editor or publisher and apparently not generally distributed, but with the statement that it is "Sold at the Society's rooms." I am indebted to A. T. Hopwood for a copy.]

Stearn, W. T. 1959. The background of Linnaeus's contributions to the nomenclature and methods of systematic biology. Systematic Zoology, 8: 4–22.

Stebbins, G. L. 1950. Variation and evolution in plants. New York, Columbia Univ. Press.

Streuli, A. 1932. Zur Frage der Artmerkmale und der Bastardierung von Baum– und Steinmarder (Martes). Zeitschr. Säugetierk., 7: 58–72.

Stroud, C. P. 1953. Factor analysis in termite systematics. Systematic Zoology, 2: 76–92.

Sylvester-Bradley, P. C. 1951. The subspecies in paleontology. Geol. Mag., 88: 88–102.

—— 1954. The superspecies. Systematic Zoology, 3: 145–146, 173.

—— ed. 1956. The species concept in paleontology. London, The Systematics Association.

Thompson, W. R. 1952. The philosophical foundation of systematics. Canadian Entomol., 84: 1–16.

Thorpe, W. H. 1940. Ecology and the future of systematics. In Huxley, 1940: 341–364.

Underwood, G. 1954. Categories of adaptation. Evolution, 8: 365–377.

Vanzolini, P. E., and L. R. Guimarães. 1955. Lice and the history of South American land mammals. Rev. Brasil. Ent., 3: 13–46.

Vicq d'Azyr, F. 1792. Système anatomique des quadrupèdes. Encyclopédie méthodique, vol. 2. Paris, Vve. Agasse.

Waagen, W. 1869. Die Formenkreise des Ammonites subradiatus. Benecke Geol.—Pal. Beitr., 2: 179–257.

Waddington, C. H. 1957. The strategy of the genes. New York, Macmillan.

Westoll, T. S. 1956. The nature of fossil species. In Sylvester-Bradley, 1956: 53–62.

White, J. A. 1953. The baculum of the chipmunks of western North America. Univ. Kansas Pub., Mus. Nat. Hist., 5: 611–631.

Wilson, E. O., and W. L. Brown, Jr. 1953. The subspecies concept. Systematic Zoology, 2: 97–111.

Wood, A. E. 1950. Porcupines, paleogeography, and parallelism. Evolution, 4: 87–98.

—— 1955. A revised classification of the rodents. Jour. Mammalogy, 36: 165–187.

—— 1958. Are there rodent suborders? Systematic Zoology, 7: 169–173.

—— 1959. Eocene radiation and phylogeny of the rodents. Evolution, 13: 354–361.

Woodger, J. H. 1937. The axiomatic method of biology. Cambridge, Cambridge Univ. Press.

—— 1945. On biological transformations. In "Essays on growth and form," W. E. Le Gros Clark and P. B. Medawar, eds.: 95–120.

—— 1948. Biological principles. London, Routledge and Kegan Paul.

—— 1952. Biology and language. Cambridge, Cambridge Univ. Press.

Zangerl, R. 1948. The methods of comparative anatomy and its contribution to the study of evolution. Evolution, 2: 351–374.

Zimmerman, W. 1953. Evolution. Die Geschichte ihrer Probleme und Erkenntnisse. Munich, Freiburg.

Zirkle, C. 1959. Species before Darwin. Proc. Amer. Phil. Soc., 103: 636–644.

Index